Essentials of
TOXIC CHEMICAL RISK *Science and Society*

Essentials of
TOXIC CHEMICAL RISK
Science and Society

Stephen Penningroth

CRC Press
Taylor & Francis Group
Boca Raton London New York

CRC Press is an imprint of the
Taylor & Francis Group, an **informa** business

CRC Press
Taylor & Francis Group
6000 Broken Sound Parkway NW, Suite 300
Boca Raton, FL 33487-2742

© 2010 by Taylor and Francis Group, LLC
CRC Press is an imprint of Taylor & Francis Group, an Informa business

No claim to original U.S. Government works

Printed in the United States of America on acid-free paper
10 9 8 7 6 5 4 3 2 1

International Standard Book Number: 978-0-415-24851-8 (Hardback)

This book contains information obtained from authentic and highly regarded sources. Reasonable efforts have been made to publish reliable data and information, but the author and publisher cannot assume responsibility for the validity of all materials or the consequences of their use. The authors and publishers have attempted to trace the copyright holders of all material reproduced in this publication and apologize to copyright holders if permission to publish in this form has not been obtained. If any copyright material has not been acknowledged please write and let us know so we may rectify in any future reprint.

Except as permitted under U.S. Copyright Law, no part of this book may be reprinted, reproduced, transmitted, or utilized in any form by any electronic, mechanical, or other means, now known or hereafter invented, including photocopying, microfilming, and recording, or in any information storage or retrieval system, without written permission from the publishers.

For permission to photocopy or use material electronically from this work, please access www.copyright.com (http://www.copyright.com/) or contact the Copyright Clearance Center, Inc. (CCC), 222 Rosewood Drive, Danvers, MA 01923, 978-750-8400. CCC is a not-for-profit organization that provides licenses and registration for a variety of users. For organizations that have been granted a photocopy license by the CCC, a separate system of payment has been arranged.

Trademark Notice: Product or corporate names may be trademarks or registered trademarks, and are used only for identification and explanation without intent to infringe.

Library of Congress Cataloging-in-Publication Data

Penningroth, Stephen.
   Essentials of toxic chemical risk : science and society / Stephen Penningroth.
      p. cm.
   Includes bibliographical references and index.
   ISBN 978-0-415-24851-8 (hardcover : alk. paper) -- ISBN 978-0-415-24852-5 (pbk. : alk. paper)
      1. Environmental toxicology. 2. Poisons--Risk assessment. 3. Science--Social aspects. I. Title.

RA1226.P46 2010
615.9'02--dc22
                                                        2009047366

Visit the Taylor & Francis Web site at
http://www.taylorandfrancis.com

and the CRC Press Web site at
http://www.crcpress.com

Dedication

This book is dedicated to the volunteers who make environmental protection a reality.

Contents

Foreword xi
Preface xiii
Acknowledgments xv
About the Author xvii

Chapter 1 Toxic Chemical Risk as Science and Social Discourse 1

    1.1   The Science of Toxicology 1
    1.2   Chemical Hazard, Risk Assessment, and Risk Management 4
    1.3   How This Book Is Organized 9
    Reference 12
    Suggested Reading 12

Chapter 2 Environmental Pathways of Toxic Chemicals 13

    2.1   Introduction 13
    2.2   Partitioning 14
        2.2.1   Evaporation 14
        2.2.2   Dissolution 16
        2.2.3   Volatilization 17
        2.2.4   Adsorption 18
    2.3   Advective Transport 19
    2.4   Chemical Transformation 21
    2.5   Bioconcentration, Bioaccumulation, and Biomagnification 24
    2.6   Ecosystems and Biogeochemical Cycles 27
    2.7   The Hydrologic Cycle 27
    2.8   Assessing and Managing Exposure 29
    References 35
    Suggested reading 35

Chapter 3 Dose–Effect: The Foundation of Toxicological Science 37

    3.1   Introduction 37
    3.2   Ethical Dilemmas and the Protection of Public Health 37
    3.3   Preliminary Investigations of Toxicity 38
    3.4   The Quantal Dose–Effect Relationship: The Workhorse of Risk Assessment 38
        3.4.1   Analysis of Incremental Toxicity 41
        3.4.2   Analysis of Cumulative Toxicity 42

	3.5	The Graded Dose–Effect Relationship 43
	Suggested Reading 52

Chapter 4 Human Populations at Risk 53

- 4.1 Introduction 53
- 4.2 Law and Loopholes 53
- 4.3 After the Fact 56
- 4.4 The Null Hypothesis and Statistical Power 57
- 4.5 Proof of Causation 58
- 4.6 Designing an Epidemiological Study: Cohort vs. Case Control 60
- 4.7 Level I, II, and III Epidemiological Studies 63
- References 66
- Suggested reading 66

Chapter 5 The Cornerstone of Risk Assessment: Toxicity Testing in Animals 67

- 5.1 Introduction 67
- 5.2 Designing a Toxicity Test 68
  - 5.2.1 Route of Exposure 69
  - 5.2.2 Dose and Time Frame 69
  - 5.2.3 Endpoint or Specified Toxic Effect 70
  - 5.2.4 Statistical Power and the Cost of Toxicity Tests 70
- 5.3 Descriptions of Toxicity Tests and Their Products 71
  - 5.3.1 Acute Lethality 71
  - 5.3.2 Subchronic Toxicity Testing 73
  - 5.3.3 Chronic Toxicity and Carcinogenicity Testing........... 73
  - 5.3.4 Reproductive Toxicity Testing 74
  - 5.3.5 Toxicity Test Design in Nonmammalian Species 75
- 5.4 The Probit Plot 76
- 5.5 Information Derived from Toxicity Testing 78
  - 5.5.1 Toxic Effect Frequencies Resulting from Specific Exposure Levels 78
  - 5.5.2 Threshold of Toxicity 79
  - 5.5.3 The Rate of Increase in Chemical Disease Frequency as a Function of Dose 81
  - 5.5.4 Potency and Efficacy 82
- 5.6 Toxicity Investigations in Individual Organisms vs. Populations 84
- 5.7 Using Animals to Screen Personal-Care Products: Local Irritation and Sensitization Tests 85
  - 5.7.1 Skin Irritation Test 85
  - 5.7.2 Eye Irritation Test 85
  - 5.7.3 Skin Sensitization Test 85

	5.8	Reducing the Use of Animals in Toxicity Testing 86
		5.8.1 Toxicity Testing in Single Cells 86
		5.8.2 Use of Bacteria to Screen Chemicals for Their Potential to Cause Cancer (Carcinogenicity) 87
		5.8.3 Use of Cultured Mammalian Cells to Screen for Genetic Toxicity 88
		5.8.4 Structure/Activity Relationships 89
	References 92
	Suggested Reading 92

Chapter 6 The Body's Defenses against Chemical Toxicity 93

    6.1 Introduction 93
    6.2 Exposure and Bioavailability 94
    6.3 The Cell Membrane 95
    6.4 Elimination by the Kidneys 98
    6.5 Excretion, Elimination, and Weak Acids and Bases 99
    6.6 Biotransformations 101
    6.7 The Kinetics of Single-Dose Exposure: Uptake, Distribution, and Elimination 107
    6.8 The Kinetics of Repeated-Dose Exposure 112
    References 118

Chapter 7 Mechanisms of Chemical Disease 119

    7.1 Introduction 119
    7.2 Noncancer Health Effects 119
        7.2.1 Organ Toxicity 120
        7.2.2 Developmental Toxicity 124
    7.3 Cancer 124
    References 134
    Suggested Reading 134

Chapter 8 Human Health Risk Assessment 135

    8.1 Introduction 135
    8.2 The Process of Risk Assessment 136
    8.3 Hazard Identification 136
    8.4 Analysis of Exposure 138
        8.4.1 Chronic Daily Intake 138
        8.4.2 Biomonitoring 141
    8.5 Analysis of Effects 142
        8.5.1 Noncancer Health Effects 142
            8.5.1.1 The Two Approaches 143
            8.5.1.2 Air Quality Index 144
            8.5.1.3 Other Strategies for Noncancer Risk Assessment 145

		8.5.2	Cancer Risk 145
		8.5.3	Risk Calculations 147
	8.6	Risk Characterization 148
		8.6.1	Uncertainty 148
		8.6.2	Weight of Evidence 149
		8.6.3	Limited Information on Chemical Toxicities 149
	References 155
	Suggested Reading 155

Chapter 9 Ecological Risk Assessment 157
    9.1   Framework for Ecological Risk Assessment 157
    9.2   The EPA's Ecological Risk-Assessment Process 158
          9.2.1   Planning 159
          9.2.2   Problem Formulation 160
          9.2.3   Analysis Phase 162
          9.2.4   Risk Characterization 165
          9.2.5   Risk Communication and Adaptive Management ... 165
    9.3   Environmental Impact Statement 166
    References 168
    Suggested Reading 168

Chapter 10 Managing Chemical Risk in North America and Europe 169
    10.1   Introduction 169
    10.2   Costs of Toxic Chemicals to Society 169
          10.2.1   Undermining Human Health 169
          10.2.2   Damage to the Environment 170
    10.3   Core Concepts of Risk Management 171
          10.3.1   Controlling Exposure 171
          10.3.2   Dealing with Uncertainty 172
          10.3.3   Navigating the Political Process 172
          10.3.4   Balancing Uncertainty, Proof, and Precaution 172
    10.4   General Strategies for Managing Toxic Chemical Risk 172
          10.4.1   Sorting Chemicals 172
          10.4.2   Limiting the Use of Toxic Chemicals 173
          10.4.3   Life-Cycle Assessment 174
    10.5   Environmental Laws in North America and Europe 174
          10.5.1   The Toxic Substances Control Act 174
          10.5.2   Canadian Environmental Protection Act (CEPA) 175
          10.5.3   REACH 177
    References 184
    Suggested Reading 185

Index 187

Foreword

Shekon (Greetings)

I hope this Foreword finds you and your families in good health and spirits.

Communities are beset by numerous problems. Pollution, contaminants and diseases that seem unique to each community are reported in the newspaper with an ever-increasing shriek. The communities respond by blaming industries, governments, and the environment, but really they have no place to go. They believe they are helpless and no one can help them. They distrust government and industry. Scientists are seen as just another expert sent in to support the power structure.

However, a community that uses its natural intelligence and resourcefulness can begin to make sense of the problems and evaluate the risks to its people and the surrounding environment. A community can solicit support and help from experts in the field; community participatory research is becoming an accepted approach to scientific investigation.

Knowing how to communicate with community members in plain language and working with their community knowledge helps to increase the effectiveness of the scientist and leads to better studies that more clearly show the risk at hand. Clear communication also helps to temper distracting rhetoric and create a context in which problems can be solved.

Better studies, increased knowledge, and clear community support result in good science and increase the credibility of the scientist working in the community. I believe that this book will help students of risk to both better understand the community and do good science.

The organization of the book aids the student of risk by weaving together three basic threads. One thread deals with the science that underlies toxic chemical risk assessment. While daunting, the text shows that the basic tools underlying these concepts are straightforward and understandable. Students with a broad background in several disciplines will find that this thread brings together much of their own knowledge, while the novice will be able to explore new understandings.

The second thread works at the smallest scale in toxicological risk assessment. The molecular basis of toxicology involves chemical keys, blockers, chains, and processes. The student of risk should have some understanding of the molecular workings of risk assessment if they are going to try to explain toxic substances to non-scientists in the community.

The last thread is the relationships among assessing toxic chemical risk, protecting human health and reducing environmental stress. This thread will lead students of risk to balance their knowledge of toxic chemicals with the practical needs of the community. Simple questions from the community take on a new level of complexity. Can we eat the fish? Can we swim in the water? These questions must be analyzed using the new tools that the student of risk has learned from this book. Many communities today are struggling with these kinds of questions. People with a working knowledge of risk are needed in these communities, and they will find that the community is their own home.

It is to be hoped that scientific knowledge and community knowledge ultimately can be welded together to produce a solid framework for understanding and managing risks from toxic chemicals. This book details basic strategies for assessing and managing risk, serving as a guide for community members who wish to become students of risk. This book can also help the scientist understand the impacts of toxic chemicals on people's lives, because it situates contamination in a community context, where the toxic effects of chemicals unfold. Scientific understanding and community well-being are, we learn, inextricably linked. Only by combining the strengths of science and community can we possibly solve these problems, not only for ourselves but for future generations.

Skennen (In Peace)

Henry Lickers
Environment Science Office
Mohawk Council of Akwesasne

Preface

This book grew out of undergraduate courses I developed and taught at Cornell University in the 1990s: "Science and Politics at Toxic Waste Sites," "Principles of Toxicology," and "Ecological Risk Assessment." The courses, which were aimed at students who did not intend to pursue a career in the field, were shaped by my experience as a technical adviser to citizens groups at three Superfund toxic waste sites in New York and New Jersey. My work as a technical adviser taught me that citizens who are concerned about contamination in their communities and who also have a basic understanding of toxicology are the single most powerful force in managing risks from toxic chemicals. This book is, in part, an attempt to help nurture informed activism at the grassroots. It is also an attempt to make the discipline of toxicology more accessible to scholars in other fields, including the social sciences and humanities, by organizing concepts and terminology that at times may seem, because of the borrowing nature of the discipline, like a confusing jumble of science, social science, and quasi science. I would like to think that overviews of basic concepts like those attempted in these pages might make it easier for others to gain a working knowledge of toxicology, especially teaching faculty at colleges and community colleges entrusted with the education of tomorrow's leaders. A more audacious hope is that this book may, in some small way, contribute to fostering change in federal laws for managing toxic chemical risk. Other countries are taking the lead with new approaches to protecting human health and the environment. America, sticking with the status quo, is in danger of falling behind. We can do better. If we value our health and the incomparable beauty of our land and waters, we will.

Acknowledgments

This book could not have been written without the support and encouragement of family, friends, and colleagues. Professor James Gillett introduced me to the discipline of toxicology during a 1987–1988 sabbatical leave at Cornell University and sponsored my first course for nontoxicologists, "Science and Politics at Toxic Waste Sites." My introduction to educational outreach came through Cornell's American Indian Program, particularly Diana Andersen, who tutored me in the practical art of interfacing community service and science, and from Henry Lickers, Ken Jock, Jim Ransom, and other members of the Akwesasne Task Force on the Environment and the tribal governments of the St. Regis Mohawk Reserve, whose battle to clean up contamination from GM and Reynolds Superfund sites bordering their territory goes on. Bob Spiegel, the executive director of the Edison Wetlands Association (EWA), continued my education in science and grassroots activism when he took a chance and hired me under a Technical Assistance Grant (TAG) from the Environmental Protection Agency (EPA). My ten-year experience as EWA's technical adviser at the Chemical Insecticide Corporation Superfund site opened my eyes to the political realities of risk management and the ways in which contamination is—and is not—cleaned up in the real world of competing stakeholder interests. I owe a special debt to Professor Rod Dietert, who sponsored "Principles of Toxicology," the Cornell course on which this book is based, and who read and critiqued the manuscript. Equally valuable were the probing questions and comments of my second reader, Nick Schipanski, who persuaded me to rethink at least a few of my assumptions. I am also grateful to my editors at Taylor & Francis for their patience in providing the long gestation time, 12 years, that I appear to have needed to write this book. But it would not have been written at all without the love and support of my wife, Judy Roberts, who kept the home fires burning—and the whole project in good-humored perspective—while I kept company with my laptop. I could not have asked for anything more.

About the Author

The son of an American newspaperman, Stephen Penningroth was educated at the Aufbaugymnasium im Schuldorf Bergstrasse in Seeheim, (West) Germany, and at Brown University in Rhode Island and the Columbia University School of General Studies in New York. He received a Ph.D. in biochemical sciences from Princeton University in 1977 and served as an assistant and associate professor of pharmacology at the University of Medicine and Dentistry of New Jersey School of Osteopathic Medicine, publishing over a dozen peer-reviewed articles on cell motility and teaching pharmacology to second-year medical and osteopathic students. Following a sabbatical leave at Cornell University in 1987–1988, he shifted focus to environmental toxicology and moved to Cornell in 1993. There he developed and taught undergraduate courses in toxicology and natural resources management while serving as a technical adviser to citizen groups at Superfund toxic waste sites. In 2000, he founded the nonprofit Community Science Institute (CSI) with a group of friends in Ithaca, NY (www.communityscience.org). He currently serves as its executive director and senior scientist. CSI operates a state-certified water testing laboratory, and its mission is to monitor water quality in the Finger Lakes region of central New York in partnership with groups of citizen volunteers and to make the results available to the general public on the Internet. The father of two grown children, Dylan and Ailey, Dr. Penningroth lives a mile from Cayuga Lake with his wife, Judy Roberts, and their two German shepherds, Xoe and Patrick.

1 Toxic Chemical Risk as Science and Social Discourse

1.1 THE SCIENCE OF TOXICOLOGY

Chemicals are double-edged swords, providing benefits to society while also posing risks. For the chemicals that society seeks to harness for its purposes, the benefits are generally easy to see and agree on. Their downsides are often less well understood.

The universe of chemicals exhibits a variety of harmful properties. Chemicals may be explosive, flammable, radioactive, corrosive, irritating, sensitizing, and/or toxic. This book addresses one harmful property, toxicity. While other harmful effects occur at the point where a chemical contacts the body, toxicity is unique in that it is almost always manifested in tissues and organs that are distant from the point of contact. Toxicology is sometimes called the "science of poisons." With few exceptions, a toxic chemical travels through the bloodstream to reach the cells it harms.

Modern toxicology is rooted in medical science, specifically in the discipline of pharmacology, where the adverse effects of medicinal chemicals have long been recognized. The discipline of toxicology has evolved over the past century to encompass not only pharmaceuticals, but metals, food additives, cosmetics, pesticides, and a host of other compounds. The growth of toxicology as a science parallels society's growing dependence on chemicals in virtually every sphere of life.

Toxicology is generally referred to as a multidisciplinary science, meaning that knowledge from other disciplines is routinely incorporated into toxicological investigations of how chemicals are transported through the environment and the body and how they produce their harmful effects. Indeed, toxicology is so multidisciplinary that it is sometimes called the "borrowing science." A partial list of disciplines from which toxicology borrows concepts and information includes medicine, pharmacology, physiology, cell biology, chemistry, molecular biology, epidemiology, earth science, ecology, agriculture, and statistics. Toxicology is also interdisciplinary: It integrates borrowed concepts and information into new patterns of knowledge specific to toxicology.

Toxicological science is concerned with what might be referred to broadly as "living systems." Examples of "living systems" include individual organisms such as a human being or a sea urchin; populations of organisms such as the people who live in a town located downwind from a power plant or coho salmon that spawn in a particular tributary of the Snake River, where water quality is threatened; and ecosystems such as a forest or a lake that comprise a variety of interacting populations of organisms, such as algae, benthic invertebrates, water fleas, and fish, together with the energy and nutrient flows that sustain them.

The term *poison* is traditionally used to describe highly toxic substances that kill at minuscule doses. In the science of toxicology, however, *poison* is a more inclusive term. All chemicals have harmful effects when the dose is sufficiently high. The relationship between the dose of a chemical and its toxic effect was first described by the fifteenth-century physician Paracelsus: "What is there that is not a poison? All things are poison and nothing without poison. Solely the dose differentiates a remedy from a poison." Paracelsus's dictum was based on his observations of medicinal drugs. However, it has been found to apply to the whole universe of chemicals, including those that have no known beneficial effects. That is, no matter what the chemical—be it sugar, table salt, Tylenol, dioxin, or lead—the greater the dose to which an organism is exposed, the greater is the likelihood of toxic effects. This pattern, which is referred to as the dose–effect (or dose-response) relationship, is the cornerstone of toxicological science.

Chemical compounds studied by toxicologists are extraordinarily diverse and resist attempts at classification. Schemes for classifying toxic chemicals do exist, but they are invariably partial, each emphasizing a particular subset of characteristics. Following are examples of categories that are sometimes employed in attempts to organize the large universe of toxic chemicals:

1. Classification based on chemical classes, e.g., polycyclic aromatic hydrocarbons, organic solvents, chlorinated hydrocarbons, and heavy metals
2. Classification based on exposure pathways, e.g., air pollutants, water pollutants, toxic chemicals in the workplace, and toxic chemicals in the home
3. Classification based on sources of toxic chemicals, e.g., plant toxins, mycotoxins (fungal toxins), industrial pollutants, and bacterial toxins
4. Classification based on how chemicals are used, e.g., food additives, therapeutic drugs, cosmetics, and agricultural pesticides
5. Classification based on mechanisms of toxicity, e.g., acetylcholinesterase inhibitors, carcinogens, neurotoxins, and reactive oxygen species

The partial nature of classification schemes is illustrated by lead, which could be assigned to any one of the listed categories. Thus, lead is a heavy metal (chemical class); it is transported through the environment via water, soil, and air (pathways of exposure); it is an industrial pollutant (source); it is used in paint, plumbing, gasoline, and military ordnance (use); and it is harmful to enzymes in cells in a variety of tissues, including the nervous system, male and female reproductive systems, blood, and kidneys (toxicity mechanisms).

The partial nature of classification schemes suggests that to understand the totality of the risks it poses, a toxic chemical is best viewed from several perspectives at once. For example, risks to human health and the environment due to lead can be evaluated on the basis of exposure to lead as an air pollutant from leaded gasoline; exposure to lead as a constituent in paint in old houses; exposure to lead in drinking water that passes through lead pipes or steel pipes containing lead solder; exposure to lead in soil at an abandoned gun factory; the chemical characteristics of lead as a positively charged metal; the mobility of lead in soil at acid pH; the toxicity of lead to the developing nervous system in children; the toxicity of lead to the female and male reproductive systems; and the toxicity of lead to red blood cells. While exceptionally multifaceted in its toxicity, lead nevertheless illustrates the basic point that risk from a toxic chemical results from a combination of chemical characteristics and the circumstances under which exposure occurs. Diverse factors may contribute to the manifestation of toxicity, such as the developmental stage of the organism that is exposed, the route by which exposure takes place, and the chemical form of the toxicant molecule when it reaches the organism. The multifactorial nature of chemical risk challenges the capacity of scientists and regulatory agencies to assess risk accurately and to manage it effectively.

Many of the factors that contribute to risk are learned from tests required by law as a condition of government approval to register and market a new chemical product. Unfortunately, due to numerous loopholes in these laws, it is estimated that only about 7% of the chemicals on the market have actually undergone comprehensive toxicity testing. Premarket toxicity testing, to the extent it is performed, has gotten better over the years since the Food and Drug Administration was created in the 1930s to serve as American society's principal gatekeeper of commercial chemicals. Yet even when premarket testing is performed, risk factors may surface after a chemical has been approved and marketed. Two examples from the 1960s and 1970s are DDT (dichlorodiphenyltrichloroethane) and thalidomide.

In the 1970s, some 40 years after its introduction as an insecticide, DDT was found to cause eggshell thinning and decreased reproduction in birds. The culprit turned out not to be DDT itself, but a chemical breakdown product, DDE (dichlorodiphenyldichloroethylene), which was generated from DDT by soil bacteria. The species at greatest risk from DDE were large birds of prey at or near the top of the food web—a consequence of two other unanticipated consequences, bioaccumulation and biomagnification: the tendency of some toxic chemicals to build up in organisms until, at the top of the food web, they reach concentrations that are thousands of times greater than their concentrations in the environment (Chapter 2.) The realization that DDT and other pesticides undergo bioaccumulation and biomagnification contributed to a ban on their use in the United States and stimulated a search for new pesticides that degrade in the environment and consequently are less likely to be magnified by food webs. (Note, however, that DDT has continued to be manufactured in the United States and exported to other countries, where it is still used in insect-control programs.)

In the 1960s, a number of babies in Europe were born with missing limbs, a result of exposure *in utero* (in the womb) to thalidomide, a drug that had recently been approved to help control morning sickness in pregnant women. Thalidomide had been subjected to toxicity testing. Why, then, had scientists failed to detect thalidomide's potential to cause birth defects? As it turned out, the toxic dose was so much higher in test animals than in humans that the effect was simply missed by the test protocols in use at the time. The teratogenicity (ability to cause birth defects) of thalidomide had been overlooked when the drug was tested in the United States as well. Approval of thalidomide had been held up by the U.S. Food and Drug

Administration for unrelated concerns—in hindsight a matter of great luck. After the thalidomide tragedy, drug testing was thoroughly reevaluated and redesigned, and it is much less likely that a teratogenic effect would be missed in prelicensing tests of a drug today.

These brief histories of thalidomide and DDT—other examples could be added to the list, such as the anti-inflammatory drug rofecoxib (Vioxx), approved in 1999 and withdrawn from the market in 2004 because of increased risk of heart attack and stroke—illustrate the toxicological complexities, scientific uncertainties, and economic factors that underlie the constant trade-offs society makes—sometimes knowingly and sometimes unwittingly—between the benefits of chemicals and the risks they pose to human health and the environment. They are also reminders that, although many risk factors can be modeled and understood on the basis of previous experience with similar chemical compounds, each chemical and the risks it poses are, in the final analysis, unique. "Compound-driven science," yet another nickname for toxicology, emphasizes the primacy of the individual chemical compound as the ultimate focus of this multidisciplinary science. In the final analysis, due to the phenomenal complexity of living systems combined with the tremendous diversity of chemicals, each individual compound can only be understood on its own terms.

1.2 CHEMICAL HAZARD, RISK ASSESSMENT, AND RISK MANAGEMENT

While a chemical may be toxic and possess other harmful properties as well, it is not considered to pose a hazard unless a human or other organism actually comes in contact with it. A chemical hazard is a composite function of a chemical's harmfulness, i.e., its toxicity, corrosiveness, radioactivity, etc., and the conditions of its use. Generally speaking, the greater the degree of contact with a harmful chemical, the greater is the hazard involved. Chemical hazard may be described qualitatively as a rectangle, with harmfulness of the chemical on one side and the degree of contact with the chemical on the other (Figure 1.1). The lengths of the two sides—intrinsic harmfulness and degree of contact—are independent of each other. The area of the rectangle symbolizes the relative hazard posed by a chemical.

Chemical hazard is not synonymous with chemical toxicity. Rather, any harmful property—explosiveness, radioactivity, corrosiveness, etc., as well as toxicity—can result in a chemical hazard if humans or other species are exposed to the chemical. Conversely, if there is judged to be no possibility of exposure, the chemical hazard is set equal to zero, regardless of how harmful the chemical may be. The dependence of chemical hazard on exposure is illustrated by Chemical C in Figure 1.1, which approaches zero as the possibility of contact between Chemical C and living organisms is thought to approach zero.

Chemical risk is synonymous with chemical hazard. Risk, like hazard, is a composite function of the harmfulness of a chemical and the conditions of its use that may result in exposure. Harmfulness and contact vary independently of each other. A chemical risk assessment is a formal evaluation of the harm a chemical can cause

Toxic Chemical Risk as Science and Social Discourse

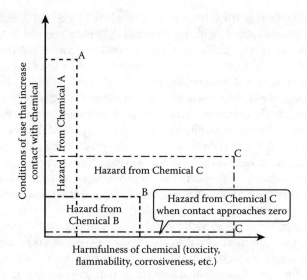

FIGURE 1.1 Chemical hazard is a composite function of the intrinsic harmfulness of a chemical and the conditions of its use. A chemical does not pose a hazard unless it is used in such a way that a harmful property is manifested through contact with living organisms. Chemical A is intrinsically less harmful than chemical B but poses a similar hazard because organisms have more contact with A than with B. The hazard from chemical C decreases to less than A or B as the probability of contact decreases, even though C is intrinsically more harmful than either A or B. (Note that the hazard rectangles in this figure are symbolic representations and are not intended to be proportional to actual chemical hazards.)

given specific conditions of use and the degree of contact that will occur. The term *chemical risk assessment* applies to an assessment of risk due to any harmful property a chemical may have. In this book, however, "chemical risk assessment" is used to refer specifically to risk from toxicity, and it is used interchangeably with "toxic chemical risk assessment."

Toxic chemical risk assessment is based on two separate analyses: an analysis of the inherent harmfulness (toxicity) of the chemical of concern, and an estimate of the degree of contact that humans or other organisms will have with the chemical (exposure). A risk assessor identifies the toxic effect or effects known to be associated with a chemical. The assessor also estimates the exposure of an at-risk population to the chemical. The magnitude of the exposure makes it possible to predict the likelihood that a particular toxic effect will occur.

Toxic chemical risk assessment rests on the dose–effect relationship, i.e., on the relationship between the dose of a chemical to which an at-risk population is exposed and the percent of individuals in the population that manifest toxicity. For all but a handful of chemicals, human data are not available. Rather, the dose–effect relationship is typically studied in laboratory animals, and the results are extrapolated to humans. The outcome, or product, of a toxic chemical risk assessment is a prediction of whether or not a population of humans or another species will suffer a specified toxic effect given the estimated level of contact with the toxic chemical. The toxic effect may be death, infertility, liver disease, blindness, reduced IQ, cancer, or some other specific illness caused by the toxic chemical. A chemical risk assessment connects exposure and toxicity. It attempts to quantitate—within the limitations and uncertainty of the scientific data—the toxic impact a given level of chemical exposure will have on an at-risk population.

A toxic chemical risk assessment may be viewed as a process for marshalling and integrating available information on the toxicity of a chemical and the exposure of an at-risk population. This process corresponds symbolically to sizing the "hazard rectangle" in Figure 1.1. While fraught with uncertainty and continually subject to revision as new scientific facts come to light, the risk-assessment process nevertheless provides a rational basis for evaluating and prioritizing risks from toxic chemicals. It is important to appreciate that risk assessment is not scientific research; rather, it is a science-based decision-making tool.

Scientists forge the decision-making tool of risk assessment, but they do not actually wield it. That role is reserved to risk managers. Risk management, like risk assessment, is a process. In contrast to risk assessment, which is a process for marshalling scientific knowledge and information, risk management is a process for making and implementing decisions about how to minimize toxic chemical risks to exposed populations. Risk-management decisions are based on science, to the extent that the toxic effects of a chemical are understood and the degree of contact can be estimated. When scientific knowledge of toxicity and exposure is incomplete—which it almost always is—risk-management decisions are influenced by factors other than science. Risk management is unavoidably a political process in which science has a preferred but by no means exclusive role in deciding how toxic chemical risks are addressed and resolved.

The political complexities of risk management arise because a broad spectrum of individuals, groups, and institutions in our society are impacted in various ways by toxic chemicals. Those who are impacted are sometimes referred to as interest groups or stakeholders. Examples of stakeholder groups are: the corporation that manufactures and profits from a chemical; workers employed by the corporation who may be exposed in the workplace; consumers who may be exposed to the chemical; federal, state, and tribal governments concerned about effects on human health and the environment; and environmental groups concerned about impacts on endangered species and ecosystems. Ideally, government brokers a negotiated agreement among stakeholder groups on how a particular toxic chemical is to be regulated, including the quantities that are manufactured; how the chemical is used; concentrations that are allowed in water, soil, and air; and procedures for disposing of the chemical safely. When the science is reasonably clear, risk-management decisions can be negotiated largely on the basis of a shared understanding of the science. But for many if not most chemical compounds in many if not most situations, the science is less than conclusive. Lack of scientific clarity has the effect of magnifying the importance of nonscience factors when stakeholders negotiate about how to manage toxic chemical risk.

Two kinds of nonscience considerations that commonly shape risk-management decisions are economic and worldview factors. Economic factors include such things as jobs that may be lost to the local economy if a manufacturing facility is closed, or a projected decrease in profitability if a manufacturer is required to spend money to reduce releases of a toxic chemical into the environment. Tradeoffs between economic factors such as these, on the one hand, and incompletely understood chemical risks, on the other, frame the political and scientific calculus of risk-management decisions.

Another set of considerations that influences chemical risk-management decisions can be described broadly as worldview factors. The term *worldview* is used here to refer to diverse decision factors that impact one's response to chemical risk after all the available scientific facts are known. Decision factors associated with worldview include such things as cultural beliefs and practices as well as perceptions of risk based on previous experience. The importance of worldview factors in risk-management decision making is suggested by two examples. In the first example, members of an American Indian tribe who engage in subsistence fishing on their reservation may attach a different meaning to a given level of PCB contamination in fish than an angler from the city who fishes a few weekends a year. Tribal members understand that they are likely to be exposed to higher levels of PCBs because they eat more fish, and the PCBs are also likely to spread through the local food web to other animals and plants that tribal members may wish to harvest as part of their diet. PCB contamination may strike at a cultural belief system such that the place where the tribe lives is seen as a gift from the Creator and its contamination by PCBs is seen as a form of desecration. In the second example, suburban homeowners may attach a great deal of significance to comparatively low levels of hazardous chemicals emanating from a recently discovered abandoned chemical waste site in their neighborhood, even though the available science indicates the health risk is negligible. They understand, first, that if the science is faulty, their children may be at risk, and second, that when it comes to real estate, a perceived health threat can lower property values as effectively as a real one. These two examples suggest how worldview factors can lead some stakeholders to manifest levels of concern that other stakeholders may view as disproportionate to the estimated chemical risk. When crafting a plan for dealing with chemical risk, government agencies need to take into account worldview factors such as cultural beliefs and economic realities in order for risk-management strategies to succeed politically.

The relative roles of science and nonscience factors in risk-management decision making are suggested in Figure 1.2. When fewer facts are available and scientific uncertainty increases, risk-management decisions are shaped increasingly by worldview and economic criteria. High decision stakes may also operate to bring nonscience criteria to the fore. Uncertainty and decision stakes (x-axis and y-axis, respectively, in Figure 1.2) vary independently of one another. The greatest source of uncertainty is often the degree of exposure of an at-risk population. The use of anthrax as an instrument of bioterrorism after September 11, 2001 illustrates this point. The lethal effects of the anthrax bacillus are well known. Managing the anthrax risk after 9/11 was complicated primarily by uncertainty about the circumstances under which people might be exposed, not uncertainty about the nature of anthrax toxicity.

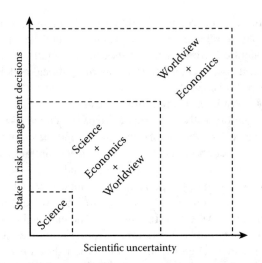

FIGURE 1.2 Risk-management decisions are based on situation-specific mixes of science, economic factors, and worldview considerations. When the scientific uncertainty and decision stakes are low, risk-management decisions are guided largely by scientific consensus. As uncertainty and decision stakes increase, scientific consensus breaks down, and risk-management decisions come to be dominated by economic and worldview factors. The unethical withholding of scientific information from public discourse or the intentional dissemination of scientific misinformation artificially increases uncertainty, undermining science-based risk management by enhancing the influence of economic, worldview, and political considerations. (Modified from Steven Rayner, Socio-cultural Definitions of Risk [Cornell University, Ithaca, NY: Institute for Comparative and Environmental Toxicology, Symposium IV, 1989], 56.)

Worldview is not the same as subjectivity or perception. The terms *subjectivity* and *perception* imply that human wishes and, indeed, human fantasies are willfully substituted for reasoned objectivity and rational debate. "Worldview," as used in this book, implies not capricious self-centeredness, but a fundamental respect for facts coupled with a recognition of honest differences in interpreting facts, fundamental limitations on human knowledge, and the unavoidable necessity of making hard decisions in the face of uncertainty. "Worldview" speaks to the political realities of real-world situations where risk-management decisions are carried out.

The ultimate goal of chemical risk management is to decide, on the basis of negotiations among stakeholder groups, the amount of risk that is acceptable when all of the available facts are known and when the diverse meanings associated with the facts are articulated by all the stakeholder groups. Risk management is not about determining and enforcing "safe" levels of chemicals in the environment. To claim that there is a "safe" level of exposure for any toxic chemical is to imply a degree of scientific certainty that simply does not exist. Experience has shown repeatedly that levels of chemicals once thought to be safe have turned out to be toxic to some at-risk populations, such as thalidomide for human fetuses, DDE for falcon fetuses, lead for young children, and Vioxx for the general population. The best that can be done in managing toxic chemical risk is to make the science as good as possible but for all stakeholders to remain aware of the limits of scientific certainty, to appreciate the importance of differences in worldview, and to be prepared to compromise. This model assumes, of course, that the political process underlying risk-management decision making is open and democratic. When facts about toxicity and exposure are known but suppressed—a classic illustration is Henrik Ibsen's nineteenth-century drama, *Enemy of the People*, about a local newspaper editor who is ostracized for trying to publicize evidence of waterborne disease in the town's new spa—important stakeholder voices are muted, the political process underlying risk management breaks down, and less powerful segments of society bear greater risk and receive fewer benefits from a chemical than others having greater power.

1.3 HOW THIS BOOK IS ORGANIZED

This book attempts to weave together three related groups of concepts. The first group, the science that underlies toxic chemical risk assessment, is presented in Chapters 2 through 5. The science of risk assessment is reasonably straightforward. It is based on empirical observations of toxicity and exposure that can be described using simple arithmetic and basic statistical concepts. Knowledge of the many scientific disciplines from which toxicology borrows is helpful but is not required to understand the principles of risk assessment.

The second group of concepts involves the physiological and molecular basis of chemical toxicity. While chemical risk assessment deals primarily with exposure and approaches at-risk organisms from outside their bodies, as it were, chemical toxicity involves studying organisms from the inside: the pathways a toxic chemical takes through the body, the organism's defenses against chemical assault, and the chemical strategies that toxic molecules use to elude the body's defenses. It is hoped that this second set of concepts, introduced in Chapters 6 and 7, may be of interest to readers with a background in the life sciences. While the molecular cell biology of toxicity is interesting to some, it is not essential to an understanding of how toxic chemical risk is assessed and managed, and these chapters may be safely omitted by readers who are not molecularly inclined.

The third group of concepts, presented in Chapters 8 through 10, describes the process of assessing toxic chemical risk to human health and the environment and the kinds of strategies that are employed in managing it. In these final chapters and throughout the book, an effort is made to relate toxic chemical risk to the real world of people and communities confronted with the vexing problem of how to deal with products that make their lives better but that also harbor a potential for harm.

STUDY QUESTIONS

1. Name three chemical hazards in addition to toxicity.
2. What distinguishes toxicity from other chemical hazards?
3. Give examples of three different levels of living systems in the hierarchy of biological organization.
4. True or false: "Some chemicals are toxic and others are not, depending on their molecular structure." Explain your answer.
5. Name three strategies for classifying toxic chemicals.

6. True or false: "The hazard presented by a chemical depends both on its intrinsic harmfulness and on how the chemical is used." Explain.
7. What is the relationship between chemical risk and chemical hazard?
8. A toxic chemical risk assessment generally involves (choose all answers that are correct):
   a. Scientific certainty (i.e., beyond a reasonable doubt) as to the types of impacts a chemical will have on an exposed population and the percent of the population that will experience the impacts
   b. Gathering and integrating information about the toxicity of a chemical and the extent to which at-risk populations have been exposed
   c. Establishing a rational basis for risk-management decisions
   d. Extrapolating toxicity data from other species to humans
   e. Calculating risks for individual people in the at-risk population based on age, gender, preexisting disease, and other factors
9. Name three broad sets of considerations that influence risk-management decisions. Does each set of considerations carry the same relative weight in every risk-management decision? Explain.
10. What is meant by "acceptable risk"? Is it the same as being "safe" from chemical toxicity? Explain.

ANSWERS TO STUDY QUESTIONS

1. Corrosiveness, explosiveness, flammability, radioactivity, irritation of the skin or eyes, and sensitization of the exposed individual to allergic (immune) reactions are six harmful properties a chemical may possess in addition to toxicity.
2. The major distinguishing feature of toxicity as a harmful chemical property is that harm is generally manifested at a site in or on the body that is physically distant from the point of contact with the chemical. To reach the site where it produces its harmful effect, a toxic chemical must enter the bloodstream and travel through the circulatory system (Chapter 6). A second distinguishing feature is that toxicity is the only universal harmful property possessed by all chemicals.
3. Examples of different levels in the hierarchy of biological organization are: an individual organism; a population of organisms of the same species; a community consisting of several populations belonging to different species; and an ecosystem comprising various communities plus sunlight, water, nutrient molecules, and other nonliving components (Chapter 2). Within an individual organism, the hierarchy of biological organization includes organs and tissues, cells, organelles, and biological molecules such as DNA and proteins (Chapter 6).
4. False. Toxicity is a universal characteristic of chemical compounds. Chemicals vary with respect to the particular kinds of toxic effects they produce and the doses at which they produce them. Beneficial chemicals such as sugar and table salt are toxic at high doses.

5. Approaches to categorizing toxic chemicals include: chemical class based on molecular structure; the pathway by which exposure takes place; the source or origin of a toxin; the context in which a chemical is used; and the mechanism by which a chemical exerts its toxic effect at the molecular level. Any chemical can be assigned to more than one category, making classification schemes partial, at best.
6. True. The hazard a chemical presents is distinct from its harmful properties. For a harmful chemical to present a hazard, it must be used or handled in such a way that humans or other species come in contact with it. If there is no contact, there is no chemical hazard, by definition. As an example, the nation's high-level radioactive waste may be stored in a permanent underground repository at Yucca Mountain in the Nevada desert. Proponents of this plan argue that the geological stability of the area guarantees zero contact with the environment. Opponents point to evidence of geological instability that might result in a rise in the water table, flooding the radioactive waste repository and potentially contaminating groundwater throughout large portions of the western United States. The hazard assessment turns on interpretations of geological data and is not cut and dried. However, it is formally correct to argue that high-level radioactive waste entombed in the Nevada desert would present zero hazard—as long as it never came in contact with the environment.
7. Chemical risk and chemical hazard are synonymous. Like chemical hazard, chemical risk is a composite function of the harmfulness of a chemical and the conditions of its use. Strictly speaking, a chemical risk may arise from any harmful property a chemical possesses (e.g., flammability, corrosiveness, toxicity, etc.). In this book, the term *chemical risk* refers to the risk associated with a chemical's toxicity.
8. Risk assessment of toxic chemicals:
    a. No. An assessment of toxic chemical risk rarely, if ever, reaches the level of scientific certainty. At best, a risk assessment may reduce the degree of uncertainty involved in a decision about managing chemical risk.
    b. Yes. Risk is a composite function of toxicity and exposure; hence risk assessment entails the gathering of information about both. Note that risk assessment always is performed for exposed populations of organisms, never for individual organisms. Risk assessment is a statistical estimate, and as such it cannot be performed for an individual organism.
    c. Yes. Establishing a rational basis for making decisions about managing risk is the sole purpose of risk assessment.
    d. Yes. Due to overriding ethical considerations, human toxicity data are rarely available. Exceptions are medicinal chemicals (drugs) and accidental poisonings by industrial and naturally occurring chemicals where human data happened to be collected. Chemical toxicity is routinely investigated in nonhuman mammalian species, and the results extrapolated to humans (Chapter 5).

e. No. Risk assessment depends on statistical analyses that apply to populations of organisms. Statistical analyses do not apply to individual organisms.
9. Three broad sets of considerations that enter into chemical risk-management decisions are: science as embodied in a chemical risk assessment; economic factors such as jobs and profitability; and worldview, including lifestyle and cultural belief systems. Every risk-management decision is site- and chemical-specific. When a risk assessment is comprehensive and the associated scientific uncertainty is low, risk management tends to be guided primarily by the science. The greater the scientific uncertainty associated with a chemical risk assessment, the greater is the role of economic and worldview factors in shaping risk-management decisions.
10. "Acceptable risk" is defined as a level of exposure to a chemical that the stakeholders in a risk-management decision collectively agree they can live with. A "safe" level of exposure is effectively impossible to define, because individual organisms in a population vary significantly with respect to their susceptibility to the toxic effects of a chemical. Further, carcinogenic (cancer-causing) chemicals can, in theory, cause cancer following exposure to a single molecule (an unlikely scenario, but theoretically not impossible; see Chapter 7 for a discussion of carcinogenicity). Finally, in the case of some chemicals, exposure levels that were once considered safe have subsequently been found to cause toxic effects in some at-risk populations (for example, thalidomide and DDT; see text in this chapter). For all these reasons, the concept of *acceptable risk* is, at bottom, a political decision arrived at by stakeholders on the basis of available science as well as economic and worldview factors.

REFERENCE

Rayner, S. 1989. Socio-cultural definitions of risk. In *Right to know*, ed. B. Lynch, 49–56. Symposium IV, Institute of Comparative and Environmental Toxicology. Ithaca, NY: Cornell University.

SUGGESTED READING

Faustman, E. M. and G. S. Omenn. 2001. Risk assessment. In *Casarett & Doull's toxicology: The basic science of poisons*, 6th ed., ed. C. D. Klaassen, 83–104. New York: McGraw-Hill.
Kamrin, M. A. 1988. *Toxicology: A primer on toxicology, principles and application*. Chelsea, MI: Lewis Publishers.

2 Environmental Pathways of Toxic Chemicals

2.1 INTRODUCTION

Toxic chemical risk is a composite function that depends both on a chemical's intrinsic toxicity and the exposure of a susceptible population to a chemical (Figure 1.1). When it comes to managing toxic chemical risk, intrinsic toxicity is a given; it cannot be changed. Exposure, on the other hand, can be reduced. Indeed, reducing exposure is the preeminent tool for managing chemical risk. Reducing exposure depends on understanding how chemicals move through the environment. The more accurately the environmental pathways of chemicals are known, the more effectively exposure and, therefore, risk can be managed.

Under what circumstances are hazardous chemicals released into the environment? One possible setting is the workplace, be it factories, agricultural fields, or office buildings. Beyond the workplace, releases generally fall into two categories, point sources and nonpoint sources. Examples of point sources include mercury emissions from the smokestack of a coal-burning power plant, runoff from a concentrated animal feedlot operation (CAFO), and discharge of wastewater from an aluminum smelter to a river. Examples of nonpoint sources include runoff from impervious urban surfaces (roofs, parking lots, streets, etc.) that may contain dirt, bacteria, petroleum products, or other chemicals; oxides of nitrogen in the exhaust of cars and trucks; and phosphorus and nitrogen in fertilizers that run off agricultural fields into streams.

Two general types of processes govern the movement of chemicals through the environment:

Diffusive transport, also called partitioning
Advective transport

Both forms of transport depend on environmental media, i.e., water, air, and soil or sediment (note: soil and sediment are treated as a single environmental medium for the purposes of this discussion). Stated differently, diffusive and advective transport depend on the abiotic environment (with one notable exception, as discussed in Section 2.5). A third process that affects chemicals in the environment, chemical transformation, plays out in the abiotic environment as well as in the biotic (living) environment, e.g., in microorganisms such as bacteria. Chemical transformation results in changes in the number of atoms in a molecule and their arrangement in space. Changes in molecular architecture can have profound effects on the diffusive transport properties of a toxic chemical. They can also alter a chemical's toxicity. Taken together, these three processes—partitioning, advective transport, and chemical transformation—determine the fate and transport of chemicals in the environment.

Two premises underlie the study of environmental fate and transport. The first premise is the conservation of mass. A chemical may move from point A in China to point B in California on prevailing westerly winds, and/or its constituent molecules may undergo rearrangement such that one or more chemically distinct compounds result, but mass itself—the sum total of the constituent atoms that comprise the original chemical—is neither created nor destroyed.

The second premise is that the fate of toxic chemicals depends on their molecular structure. Many toxic chemicals are transformed (broken down or metabolized) into their molecular building blocks and recycled in the environment. Some, referred to as *persistent organic pollutants* (POPs), continue unchanged in the environment for months, years, or even decades. While POPs can be broken down and recycled eventually, at least in principle, the same is not true of toxic metals such as mercury and lead. Metals are elements, and as such they cannot be broken down. Once introduced into the environment, they persist indefinitely.

In summary, the movement of a chemical through the environment follows a set of pathways, each one resulting from a unique combination of pure chance with the laws of physics, chemistry, and biochemistry. Exposure occurs when a population of organisms happens to serve as a way station on one of the environmental pathways taken by a chemical or its transformation products. Gauging these pathways begins with an introduction to partitioning and advective transport.

2.2 PARTITIONING

Matter exists in various physical-chemical states. Depending on its temperature, it may exist as a solid, a liquid, or a gas. In addition, it may dissolve in water or other liquid. Molecules in solution may adsorb to a surface with which they come in contact.

Partitioning, or diffusive transport, is a spontaneous passive process whereby the molecules comprising a chemical distribute themselves between two physical-chemical states by diffusion. For example, water molecules in a glass partition themselves between the water and the air above the glass. Phosphorus molecules dissolved in shallow subsurface waters partition themselves between the water and adsorbing to the surface of soil particles. A chemical's intrinsic properties determine the extent to which it distributes itself between two physical-chemical states. Evaporation, dissolution, volatilization, and adsorption are partitioning processes that have important roles in the environmental fate and transport of chemicals.

2.2.1 Evaporation

Diffusive transport by evaporation involves a change in the bulk physical state of a chemical. When a chemical such as water or benzene is in a liquid state, individual molecules continuously and spontaneously detach themselves from the liquid and enter the air as water vapor or benzene vapor, i.e., as a gas. Conversely, vapor molecules continually condense into liquid. One way to study the energy involved in the transition between liquid and gas is to place the liquid in a closed container under standard conditions and wait until the liquid chemical and its vapor come into equilibrium. At equilibrium, molecules continue to evaporate from the liquid into the air space in the closed container, and gas molecules condense, leave the air space, and reenter the liquid. Because the rate of movement in each direction remains constant, the ratio of the mass of chemical in the gaseous state to the mass of chemical in the liquid state is also constant. The pressure exerted by a gas above a liquid in a closed container at equilibrium under standard laboratory conditions is referred to as the liquid's vapor pressure. The units of vapor pressure are millimeters of mercury (mm Hg). As might be imagined, the energy of heat drives the evaporation reaction, and vapor pressure increases and decreases with temperature.

The vapor pressures of several organic chemicals are listed in Table 2.1. Comparing vapor pressures makes it possible to predict the relative tendencies of liquids to evaporate. Let's assume, for example, that pure benzene, pure dioxin, and pure vinyl chloride are discovered in three separate puddles at a hazardous chemical waste site. Based on the vapor pressures listed in Table 2.1, 132 billion times more benzene should evaporate than dioxin, while 27.9 times more vinyl chloride should evaporate than benzene. The actual amounts that enter the atmosphere depend, of course, on how big the puddles are. If the benzene puddle is not 132 billion times larger than the dioxin puddle, it will not contribute 132 billion times more benzene. Other "real-world" factors may also affect the relative amounts of chemical that evaporates. Vapor pressures are determined in closed containers under standard conditions of temperature and pressure, yet at a typical hazardous chemical waste site, temperatures change continuously, and chemicals are exposed to the weather. In complex open systems such as these, equilibrium between the liquid and gaseous states of a chemical is never reached. Thus, vapor pressures determined under

TABLE 2.1
Vapor Pressures and Henry's Law Constants of Selected Organic Chemicals

Chemical	Vapor Pressure (mm Hg)	Henry's Law Constant (atm×m³ mol⁻¹)
Benzene (C_6H_6)	95.2	5.56×10^{-3}
Phenol (C_6H_6O)	62×10^{-2}	2.7×10^{-7}
Pentachlorophenol (C_6HCl_5O)	7.5×10^{-5}	2.8×10^{-7}
2-Nitrophenol ($C_6H_5NO_3$)	8.9×10^{-2}	3.5×10^{-6}
Dioxin ($C_{12}H_4Cl_4O_2$)	7.2×10^{-10}	5.4×10^{-23}
Vinyl chloride (C_2H_3Cl)	2,660	2.65×10^{-2}
Trichloroethylene (C_2HCl_3)	69	7.69×10^{-3}
Malathion ($C_{10}H_{19}O_6PS_2$)	35.3×10^{-6}	4.89×10^{-9}
Lindane ($C_6H_6Cl_6$)	802.5×10^{-6}	2.43×10^{-7}

Source: Data from John H. Montgomery, *Groundwater Chemicals Desk Reference*, 3rd ed. (Boca Raton: CRC Press, 2000.)

standard conditions such as those in Table 2.1 do not provide an exact measure of how much liquid enters the air as a gas. However, they can be used to obtain ballpark estimates, or educated guesses, of the amount of a liquid that enters the air by evaporation over time.

2.2.2 Dissolution

Like evaporation, dissolution involves the transition of a bulk chemical to a molecular dispersion. While evaporating molecules disperse in air, dissolving molecules disperse in a liquid. For example, sugar dissolves in a cup of coffee, and alcohol dissolves in a bottle of wine. The tendency of a bulk chemical to dissolve in water is referred to as its aqueous solubility, which is defined as the mass of a chemical that dissolves in water under standard conditions of temperature and pressure. The units of aqueous solubility are typically milligrams of chemical dissolved per liter of water (mg L^{-1}). The aqueous solubilities of selected organic chemicals are listed in Table 2.2. Inorganic chemicals such as metals and salts tend to be much more water-soluble than organic chemicals. From the standpoint of toxic chemical risk, the partitioning process of dissolution, like that of evaporation, results in a hazardous chemical diffusing from a small space where it is present in bulk form into an environmental medium where it can become more widely distributed.

In addition to aqueous solubility, a second approach is often used to measure the partitioning process of dissolution. If you've ever mixed oil with water—say, automotive oil or cooking oil—and then watched as the oil and water separated after you stopped shaking, you've observed the inspiration for this second approach: Rather than measuring how well a chemical dissolves in water, compare instead how well a chemical dissolves in water, on the one hand, and in a second liquid that does not

TABLE 2.2
Aqueous Solubilities, Octanol:Water Partition Coefficients, and Soil:Water Partition Coefficients of Selected Organic Chemicals

Chemical	Aqueous Solubility (mg L^{-1})	Octanol:Water Partition Coefficient (log K_{ow})	Soil Sorption Coefficient (log K_{oc})
Benzene (C_6H_6)	1,740	2.13	1.69
Phenol (C_6H_6O)	84,045	1.57	2.7
Pentachlorophenol (C_6HCl_5O)	21.4	4.07	3.41
2-Nitrophenol ($C_6H_5NO_3$)	1,300	1.62	2.96
Dioxin ($C_{12}H_4Cl_4O_2$)	0.00032	6.2	6.6
Vinyl chloride (C_2H_3Cl)	1,100	0.6	0.39
Trichloroethylene (C_2HCl_3)	1,100	2.27	1.81
Malathion ($C_{10}H_{19}O_6PS_2$)	145	2.84	2.61
Lindane ($C_6H_6Cl_6$)	9.2	3.89	2.86

Source: Data from John H. Montgomery, *Groundwater Chemicals Desk Reference*, 3rd ed. (Boca Raton: CRC Press, 2000.)

mix with water, on the other. Octanol, a straight-chain hydrocarbon with eight carbon atoms, is an organic solvent that is commonly used to compare solubilities. The test chemical is added to a mixture of water and octanol, the mixture is shaken vigorously, and the water and octanol are allowed to separate into two distinct phases, with the octanol on top and the denser water on the bottom. The concentration of the test chemical is determined in each layer. The ratio of the concentrations of the chemical in octanol and in water is a measure of the tendency, or "preference," of the chemical to dissolve in an organic solvent rather than in water. The logarithm of the concentration ratio is called the octanol:water partition coefficient, or K_{ow}. The octanol:water partition coefficient values of a few compounds are listed in Table 2.2. The octanol:water partition coefficient of a chemical tends to be inversely related to its aqueous solubility: The higher the octanol:water partition coefficient, the lower is the aqueous solubility.

2.2.3 Volatilization

In the partitioning processes of evaporation and dissolution, chemicals go from bulk forms to molecular dispersions in air and water. The partitioning process of volatilization begins with chemicals that are already dissolved in a liquid such as water. Some of the dissolved molecules volatilize, i.e., they spontaneously exit the water and enter the surrounding air. The rate of volatilization depends on the molecular structure and intrinsic properties of the chemical. Like all chemical reactions, volatilization is a two-way street: At the same time molecules in water volatilize into the air, molecules in the air enter and dissolve in water. The tendency of a chemical to distribute itself between a dissolved state in water and a gaseous state in air is defined as the ratio of a chemical's concentration in air to its concentration in water in a closed container under standard conditions:

$$C_g/C_w = H \tag{2.1}$$

where C_g is the concentration of chemical in the gas phase, usually expressed in units of atmospheres (atm); C_w is the concentration of chemical in water, usually expressed in units of moles of chemical per cubic meter (mol m^{-3}); and H is Henry's law constant in units of atm×m^3 mol^{-1}. (C_g and C_w may also be expressed in units of mg L^{-1}, in which case Henry's constant is dimensionless.) The Henry's law constants of selected organic compounds are listed in Table 2.1. Note that unlike K_{ow}, which is the logarithm of a concentration ratio, H is a concentration ratio, not its logarithm.

The values in Table 2.1 suggest a general correlation between vapor pressures and Henry's law constants, i.e., compounds that have similar vapor pressures also tend to have similar Henry's law constants. However, there are many exceptions. For example, phenol and pentachlorophenol have similar Henry's law constants, but their vapor pressures differ by a factor of 10,000. In other words, phenol and pentachlorophenol volatilize at similar rates, but they evaporate at vastly different rates. Volatilization and evaporation are different physical chemical processes with different energy requirements. A volatilizing molecule must overcome forces that hold it in solution, such as its attraction to water. An evaporating molecule must overcome forces that hold it in its bulk form, such as its attraction to other molecules like itself. These forces may be quite different, resulting in different rates of volatilization and evaporation for related chemicals.

As noted previously, partitioning by volatilization is a two-way street: There is a continuous back-and-forth between water and air, in keeping with the bidirectionality of all chemical processes. With respect to the chemical process of volatilization, the net movement of chemical between water and air depends on its Henry's law constant as well as on its relative concentrations in water and air; the dependence on relative concentrations is an example of the law of mass action. At equilibrium, the net movement of molecules ceases, but the molecules themselves do not stop moving. Rather, they continue to move at a constant rate from water to air and at another, constant rate from air to water. The two rates at which molecules move are different at equilibrium, but they are constant.

While the process of volatilization can be an important factor in the fate and transport of hazardous chemicals in the environment, it is also essential to life. To give two examples: The concentration of oxygen in a stream, a basic requirement for fish and other aquatic life, is the net result of oxygen from the atmosphere dissolving in the stream and the dissolved oxygen volatilizing back into the atmosphere. The removal of metabolic wastes from our bodies depends on carbon dioxide volatilizing from pulmonary capillary blood into air to be exhaled from our lungs at the same time that oxygen in inhaled air is dissolving in capillary blood. Volatilization is a fundamental physical-chemical process without which life would not be possible.

2.2.4 Adsorption

Volatilization defines the partitioning of a chemical between water and air. Adsorption defines the partitioning of a chemical between water and soil. In the process of adsorption, which is also referred to variously as sorption or retention, molecules move back and forth between being dissolved in water and being attached to the surfaces of soil or sediment particles with which the water is in contact. How a chemical distributes itself between being adsorbed to soil and dissolved in water is described by the adsorption coefficient, or the soil:water partition coefficient. The ratio of the concentrations of adsorbed to dissolved chemical at equilibrium under standard conditions is:

$$C_s/C_w = K \qquad (2.2)$$

where C_s is the concentration of chemical adsorbed to soil in units of mg kg^{-1}, C_w is the concentration of chemical dissolved in water in units of mg L^{-1}, and K is the soil:water partition coefficient (also called the adsorption coefficient) in units of L kg^{-1}. Because it is a ratio between the concentrations of a chemical in two environmental media that are in contact with each other, the adsorption coefficient is analogous to Henry's constant (Equation 2.1). The adsorption coefficients of selected chemicals are presented in Table 2.2. Note that the adsorption coefficient is referred to as K_{oc} because it is normalized to the organic content of soil. Note, too, that like the octanol:water partition coefficient, K_{ow}, it is reported in the form of its logarithm.

A wide variety of chemicals, including charged ions like arsenic and lead as well as carbon-rich organic compounds such as pesticides and PCBs (polychlorinated biphenyls), tend to move from a state of dissolution in water to a state of attachment to soil. Why are diverse chemical compounds attracted to soil? Because soil is heterogeneous with respect to its composition and polarity. A soil particle may have negative or positive electrical charges in some places on its surface while lacking electrical charges in others. Where a soil particle has a patch of exposed electrical charges, ions of opposite charge (polar molecules) may be sufficiently attracted to leave their dissolved state in water and adsorb to the charged patch. While nonpolar molecules are indifferent to electrical charge, they may "prefer" adsorption to uncharged surface patches to their dissolved state in water. Adsorption based on either polar or nonpolar forces can be quite strong and result in a large soil:water partition coefficient.

Like volatilization, adsorption is a dynamic, two-way process in which molecules move back and forth between two environmental media. Whether the net movement of a chemical is from soil to water or water to soil is determined, again, by two factors: the soil:water partition coefficient, and the relative concentrations of the chemical in soil and water. If the chemical's soil:water concentration ratio exceeds the soil:water partition coefficient, the net movement of chemical is from soil to water. If the soil:water concentration ratio is less than the soil:water partition coefficient, the net movement is from water to soil. When net movement ceases, molecules continue to move, but at a constant rate in each direction.

The soil:water partition coefficient is reduced by the presence in water of substances that decrease a chemical's binding to soil or increase its solubility in water. For example, positively charged lead, Pb^{2+}, or mercury, Hg^{2+}, ions bind less tightly to negatively charged sites on soil particles when water contains elevated concentrations of positively charged hydrogen ions, H^+ (e.g., from acid), or sodium ions, Na^+ (e.g., from salt). Positive ions such as H^+ and Na^+ compete with Pb^{2+}, Hg^{2+}, and other positive ions for binding to negatively charged sites on soil particles, effectively lowering their soil:water partition coefficients. The soil:water partition coefficient of nonpolar chemicals such as pesticides and PCBs may be lowered by adding an organic solvent such as acetone, which increases their solubility in water. Compounds that are miscible with water and enhance the solubility of nonpolar compounds are said to act as cosolvents.

2.3 ADVECTIVE TRANSPORT

Whether a toxic chemical enters the environment by partitioning into water, soil, or air has profound consequences for exposure. Different environmental media are subject to different physical forces. Chemicals that have partitioned into a particular environmental medium are like hitchhikers that have no choice but to go wherever their "ride" takes them. Each "ride" in an environmental medium is a form of what is referred to as *advective transport*.

In advective transport, chemicals that have partitioned into water, soil, or air now move with the environmental medium. Chemicals dissolved in water follow its downhill path, gaseous chemicals follow air currents, and adsorbed chemicals accompany soil and sediment particles on their journey. The forms of advective transport are as diverse as environmental media and the forces that make them mobile. Take, for example, a hypothetical chemical A that has partitioned by dissolving in surface water. The movement of surface runoff to a stream is a form of advective transport of dissolved chemical A. Hypothetical chemical B is adsorbed to soil particles. Surface runoff containing these soil particles is a form of advective transport for chemical B. When soil particles that are not washed into the stream dry out and blow away, chemical B is being advectively transported by air currents. A stream receiving runoff may carry dissolved chemical A and adsorbed chemical B to a lake, in which case the stream will have advectively transported chemicals A and B. Soil particles falling to the bottom of the lake with chemical B still attached represent another stage of advective transport. When surface water containing dissolved chemical A percolates downward through the soil, it advectively transports chemical A into groundwater. Chemical A then ends up "hitching a ride" with groundwater as it flows to the nearest stream, lake, or other surface-water body.

Even if a chemical does not partition into an environmental medium, it may nevertheless undergo advective transport. For example, oil spilled at sea dissolves poorly in water, and most of it floats on the water's surface. The undissolved oil is transported advectively by the water, its pathways determined by wind, wave action, and ocean currents. Groundwater, too, may transport undissolved organic chemicals advectively, e.g., trichloroethylene that was spilled on the ground at an industrial site and percolated into a groundwater aquifer beneath the site.

To manage exposure, one often needs to evaluate the mass of a hazardous chemical moving from a source to a point of contact with a receptor via advective transport. (Note: In the context of assessing exposure and risk, *receptor* refers to any biological or ecological entity that may be impacted by a toxic chemical [Chapter 9].) The mass of advectively transported chemical can be calculated if three parameters are known

1. The concentration of the chemical in the environmental medium
2. The rate at which the environmental medium moves through the landscape
3. The time over which the movement takes place

Consider, for example, the amount of phosphorus that is transported to a lake by a stream that runs through agricultural cropland. Phosphate and nitrate are essential plant nutrients. At elevated concentrations, however, they act as pollutants by promoting excessive growth of algae and other plants that can change a lake's ecosystem. The Chesapeake Bay is an example of a water body that has been damaged by excessive nutrients as well as sediment from as far upstream as the Susquehanna River in New York State. Various programs attempt to manage agricultural and other sources of nutrients to reduce harmful impacts on downstream water bodies. One measure of effectiveness is the mass (amount) of nutrient that is transported advectively by a stream. For example, the amount of phosphorus that is transported, which is also referred to as the *load*, is determined in three steps:

1. Measure the concentration of phosphorus in the stream at a point near its confluence with the lake.
2. Measure the flow of the stream at the same point where the phosphorus concentration is measured, i.e., the volume of water that flows past the point per unit time.
3. Multiply the concentration of phosphorus times the flow of the water to get the phosphorus load per unit time, e.g., milligrams per second, grams per day, kilograms per month, or tons per year.

Determining the phosphorus load is simple in concept. However, concentration and flow vary from day to day, and the financial resources to measure them accurately over time are seldom available. As a result, loadings are almost always estimates, not precise calculations. The accuracy of a loading estimate depends on the degree to which concentration and flow measurements are representative of the stream as well as the quality of the model used to extrapolate estimates to the majority of days when no samples can be taken.

In principle, the advective transport of any chemical by any environmental medium may be estimated using the same approach as the load of phosphorus in a stream: chemical concentration in the environmental medium, rate of movement of the environmental medium through the landscape, and the time period over which the movement takes place. The accuracy of the determination depends, as with stream loads, on the accuracy with which these parameters can be measured. Performing accurate estimates of advective transport can present formidable challenges, even for order-of-magnitude estimates that generally suffice for risk-management purposes.

2.4 CHEMICAL TRANSFORMATION

In addition to partitioning and advective transport, chemicals that enter the environment are subject to structural transformations that can affect the pathways they take. *Transformation* means a structural change at the level of the individual molecules that make up a chemical: a change in the number of atoms in the molecule and how they are arranged in space. Changes in molecular structure can affect partitioning. They may also decrease the toxicity of a chemical or, occasionally, cause an increase in toxicity.

Chemical transformations may result from abiotic forces, e.g., water, sunlight, ultraviolet radiation, or contact with other chemicals, as well as biotic forces, i.e., action by living organisms, usually microbes such as bacteria and fungi. One example of abiotic chemical transformation (actually, a series of abiotic chemical transformations) is the generation of sulfuric acid, one of the main constituents of acid rain, from the sulfur present in coal used to fuel electric power plants. Sulfur (S) present in coal is oxidized by oxygen in the air (O_2) to produce sulfur dioxide (SO_2) as coal is burned at high temperatures. After being emitted by the power plant to the atmosphere, the sulfur dioxide is oxidized further to sulfur trioxide (SO_3). Sulfur trioxide, in turn, reacts with water vapor in the atmosphere (H_2O) to form sulfuric acid (H_2SO_4). The sequence of abiotic transformation reactions is:

$$S + O_2 \leftrightarrow SO_2$$

$$SO_2 + 1/2\ O_2 \leftrightarrow SO_3$$

$$SO_3 + H_2O \leftrightarrow H_2SO_4$$

The sulfuric acid is generated in atmospheric water droplets and remains in solution, undergoing advective transport with the droplets on air currents. Eventually, some of the acidified water droplets fall to the earth as acid rain. In the United States, coal-burning power plants in the Midwest have been a source of acid rain in the East, and this was a major impetus behind passage of the Clean Air Act in 1990.

An example of biotic chemical transformation is the conversion of elemental mercury (Hg) to methylmercury ($HgCH_3$):

$$Hg + CH_3 \leftrightarrow HgCH_3$$

The chemical transformation of elemental mercury to methylmercury can be performed by bacteria in water and soil. Elemental mercury and methylmercury are both toxic to the central nervous system. Elemental mercury evaporates, and the toxic vapor is inhaled. The saying "mad as a hatter" arose from the occupational exposure of hatters to vapors from liquid mercury, which was used for centuries in the manufacture of felt hats. While mercury vapor is very toxic, the liquid metal itself is less so. It does not dissolve well in water, and its poor solubility tends to limit its access to living systems. Methylmercury, on the other hand, dissolves in water and readily enters the body by all exposure routes: from the intestinal tract following ingestion, through the skin, and through the lungs. Because exposure to methylmercury can happen quickly and easily, it is considered to be an extremely hazardous chemical.

A tragic incident in Minamata, Japan, in the 1950s made clear the health risks posed by methylmercury in the environment. An acetaldehyde plant discharged mercury and methylmercury into Minamata Bay; the mercury in the discharge is believed to have been transformed to methylmercury by aquatic bacteria, adding to the concentration of methylmercury. Methylmercury entered the food web of Minamata Bay and underwent biomagnification, resulting in high concentrations in fish and shellfish (see Section 2.5). Humans were exposed when they ate seafood that was caught and sold locally. Adults suffered a wide range of neurological disorders. Babies exposed *in utero* were born with crippling birth defects. Minamata disease, as it came to be called, suggests how the biosphere can contribute to toxic chemical risks as well as be impacted by them. In the case of Minamata disease, exposure occurred, in part, because aquatic bacteria transformed mercury to methylmercury, and because methylmercury was advectively transported by fish and other seafood from the water to humans.

Environmental Pathways of Toxic Chemicals

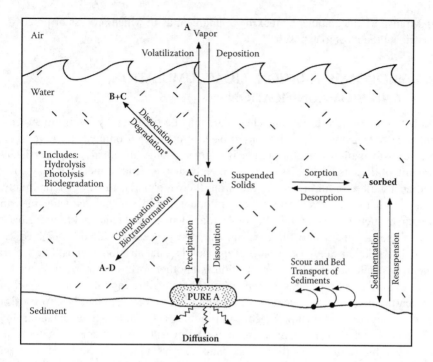

FIGURE 2.1 Schematic representation of partitioning, advective transport, and chemical transformation processes in an aquatic environment. Depicted are the dissolution-precipitation, volatilization-deposition, and sorption-desorption (two-way partitioning processes) of chemical A; sedimentation of chemical A adsorbed to soil/sediment particles and their scouring (i.e., downstream transport) by the stream current (advective transport); and the transformation of dissolved chemical A to degradation products B and C by means of chemical reactions that are abiotic (e.g., hydrolysis, photolysis) or biotic (biotransformation). Biotransformation reactions by organisms such as aquatic bacteria may be facilitated by large biological molecules such as D, which complex with A and present it to the bacterial cell for processing. (Reprinted with permission from Gary Rand, ed., *Fundamentals of Aquatic Toxicology*, 2nd ed. [Washington, D.C.: Taylor and Francis, 1995], 450.)

The diagram in Figure 2.1 illustrates a number of fate and transport processes that operate on a hypothetical pure chemical solid called "A" that has been placed on the bottom of a stream. Chemical A partitions into the water by dissolving. Once dissolved, chemical A may undergo further partitioning by adsorbing to suspended sediment particles or volatilizing into the air. Chemical A may undergo chemical transformation to daughter compounds B and C as a result of chemical reactions caused by abiotic processes (hydrolysis by water, photolysis by sunlight) or biotic processes (uptake and biodegradation by microorganisms). Biodegradation may entail chemical A first associating with a large biological molecule, D, which facilitates A's transformation. Chemical A and its transformation products B and C are subject to advective transport by water and air. Figure 2.1 illustrates the complexity of environmental fate and transport, in that multiple processes typically take place simultaneously. Risk assessments often depend on simplifying environmental fate and transport calculations by making a judgment as to which process or processes contribute the most to exposure.

2.5 BIOCONCENTRATION, BIOACCUMULATION, AND BIOMAGNIFICATION

As noted previously, hazardous chemicals may be taken up by living organisms, chemically transformed, and released back into the environment. However, neither transformation nor elimination is a given. Some chemicals, e.g., persistent organic pollutants like PCBs or heavy metals such as lead or mercury, may enter and accumulate in an organism, retaining their molecular identities indefinitely. When a chemical moves from an environmental medium such as water into an organism such as a fish, retains its molecular identity, and tends not to move back out of the organism into the environment, the process is called *bioconcentration*. The body has defenses that prevent the bioconcentration of most but not all chemicals (Chapter 6). Bioconcentration is another form of partitioning, analogous to volatilization or adsorption, except that partitioning takes place between an abiotic environment (such as water) and a biotic environment (such as a fish or other aquatic organism). Bioconcentration is a ubiquitous phenomenon, and it can occur when soil organisms such as earthworms contact contaminated soil, when benthic (bottom-dwelling) organisms such as aquatic insect larvae contact contaminated sediment, and when birds or mammals contact contaminated air. In effect, the biosphere behaves like a fourth environmental medium—in addition to water, air, and soil/sediment—into which some chemicals are distributed by diffusive transport.

Most hazardous chemicals do not build up inside the organisms they enter. The reason is that living organisms have evolved effective systems for eliminating most undesirable chemicals from their bodies by a combination of chemical transformation and other mechanisms (Chapter 6). Bioconcentration results when a xenobiotic (a chemical that is foreign to the body; from the Greek *xeno*, meaning foreign) enters an organism at a faster rate than the organism is equipped to eliminate it.

The ratio of the concentration of a hazardous chemical in an organism to its concentration in an environmental medium is referred to as the bioconcentration factor:

$$C_o/C_{em} = BCF \qquad (2.3)$$

where C_o is the concentration of chemical in the organism, C_{em} is the concentration in the environmental medium, and BCF is the bioconcentration factor under standard laboratory conditions. Typical units for BCF are liters per kilogram (L kg^{-1}). Because it is based on a passive partitioning process, the BCF is analogous to the soil:water partition coefficient, K_{oc} (Equation [2.2] and Table 2.2); Henry's constant, H (Equation [2.1] and Table 2.1); and the octanol:water partition coefficient, K_{ow} (Table 2.2).

The BCF can be used to estimate the concentration of a chemical in an exposed population of organisms when the chemical's concentration in the environmental medium is known. The bioconcentration factors of selected chemicals in fish and algae are listed in Table 2.3. As a general rule of thumb, the bioconcentration factors of organic chemicals parallel their octanol:water partition coefficients and soil:water partition coefficients (compare Tables 2.3 and 2.2). Still, bioconcentration may be affected by species differences. For example, the bioconcentration factors for cadmium differ by more than a factor of 100 between two species of fish (Table 2.3).

Bioconcentration refers to the passive partitioning of a xenobiotic between an abiotic environmental medium and a living organism. Bioaccumulation describes the uptake and retention of a xenobiotic from all sources, biotic (prey) as well as abiotic. Bioaccumulation depends on the same phenomenon as bioconcentration: The rate at which an organism takes up a xenobiotic exceeds the rate at which the organism is able to eliminate it. As a result, more chemical enters the body than leaves it per unit time.

Every ecosystem is characterized by a complex and interconnected hierarchy of predator-prey relationships called a *food web* (Figure 2.2). Biomagnification is a phenomenon whereby species at higher levels of the food web, such as hawks, whales, and humans, are found to have higher concentrations of bioaccumulative chemicals than species at lower feeding levels, such as algae and water fleas. Biomagnification results from specific sequences of predator-prey relationships, called *food chains*, within the larger food web. At low feeding levels, species such as algae bioaccumulate a chemical primarily or exclusively through passive bioconcentration from an environmental medium such as water. When predators at successively higher trophic levels feed selectively on contaminated prey species—e.g., zooplankton feed on algae,

TABLE 2.3
Bioconcentration Factors of Selected Organic Chemicals and Cadmium

Chemical	Bioconcentration Factor (log BCF)	
	Fish[a]	Algae[b]
Benzene (C_6H_6)	0.63	1.48
Phenol (C_6H_6O)	1.24	0.54
Pentachlorophenol (C_6HCl_5O)	2.89	3.10
2-Nitrophenol ($C_6H_5NO_3$)		1.48
Dioxin ($C_{12}H_4Cl_4O_2$)	5.8	
Vinyl chloride (C_2H_3Cl)		1.6
Trichloroethylene (C_2HCl_3)	1.23	3.06
Malathion ($C_{10}H_{19}O_6PS_2$)	1.58	
Lindane ($C_6H_6Cl_6$)	2.65	2.38
Cadmium (Cd)	1.52[c]	3.41
	3.61[d]	

Source: Data from John H. Montgomery, *Groundwater Chemicals Desk Reference*, 3rd ed. (Boca Raton: CRC Press, 2000); and Ronald Eisler, *Handbook of Chemical Risk Assessment, Vol. I. Metals* (Boca Raton: Lewis Publishers, 2000.)

[a] Various species of fish.
[b] Various species of algae.
[c] *Oncorhynchus mykiss.*
[d] *Gambusia affinis.*

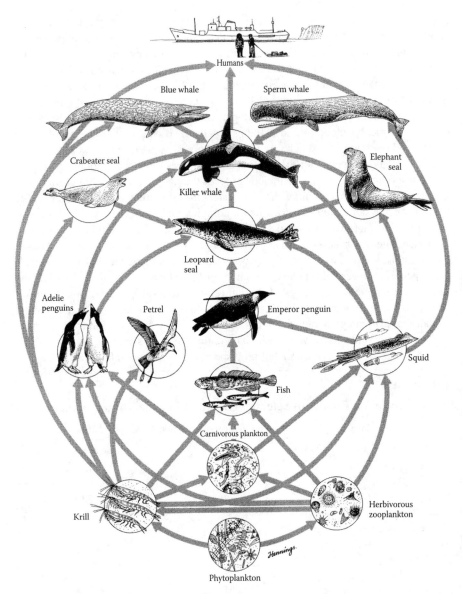

FIGURE 2.2 Greatly simplified food web in the Antarctic. There are many more participants in the food web that are not shown, including an array of decomposer organisms. (Reprinted with permission from G. Tyler Miller, Jr., *Environmental Science*, 3rd ed. [Belmont: Wadsworth Publishing, 1991], 73.)

fish feed on zooplankton, and ospreys feed on fish—the bioaccumulative effect of the toxic chemical is magnified, and concentrations at the top of the food chain can reach levels thousands of times greater than those at the bottom. Biomagnification is essentially a special case of bioaccumulation. Minamata disease resulted from the biomagnification of methylmercury through the food web in Minamata Bay.

The tragic birth defects of Minamata disease illustrate a sad truth: The young of all top predator species, not just humans, are uniquely vulnerable to chemical assault through biomagnification. Species differ dramatically in many ways, but when it comes to cellular defenses against bioaccumulative chemicals, especially during the critical early stages of development in the womb or the egg, they are remarkably similar. Their shared vulnerability offers toxicological confirmation of the fundamental relatedness of diverse species.

2.6 ECOSYSTEMS AND BIOGEOCHEMICAL CYCLES

Food webs that can act to magnify the effects of bioaccumulation are embedded in ecosystems. A detailed introduction to ecosystems is beyond the scope of this book. However, a brief overview may illuminate the general ecological context in which biomagnification occurs.

An ecosystem may be loosely defined as a self-sustaining assemblage of biotic and abiotic parts. The assemblage is shaped by geography and landscape. Two broad classes of ecosystems are aquatic, such as a lake or stream, and terrestrial, such as a forest or field. The cornerstone of an ecosystem—indeed of life on earth—are green plants, which are referred to as producers, or autotrophs (from the Greek *auto,* meaning self, and *troph,* meaning feeder). Plants are able to grow by utilizing energy from the sun and inorganic nutrients from soil, water, and air in a complex set of biochemical reactions known as photosynthesis. Photosynthesis has two momentous outcomes for life on earth: the incorporation of carbon from the environment into the body of the plant, which animals can then feed on; and the production of oxygen from water, which animals can breathe. The bodies of green plants are the foundation on which every ecosystem's food web is built.

The animals in an ecosystem are called consumers, or heterotrophs (from the Greek *hetero,* meaning other, and *troph,* meaning feeder), because animals are completely dependent on plants and/or other animals as their food sources. Primary consumers are herbivores that feed only on plants. Secondary consumers are carnivores that feed on herbivores. Tertiary consumers are carnivores that feed on other carnivores. Producers and consumers occupy successive trophic levels in food chains. An ecosystem's food web consists of diverse overlapping and interlocking food chains (Figure 2.2). Food webs include decomposer organisms that return abiotic chemicals such as carbon, nitrogen, and phosphorus from dead organisms to their respective biogeochemical cycles, whence they can be accessed, once again, by autotrophic plants.

2.7 THE HYDROLOGIC CYCLE

Biogeochemistry is the study of the processes that govern the chemical composition of the environment, including flows of matter and energy. The food webs of ecosystems depend on biogeochemical cycles of carbon, nitrogen, phosphorus, and water. These cycles provide plants with nutrients and operate on interlocking local, regional, continental, and global scales. The most familiar of the plant nutrient cycles is the hydrologic cycle (Figure 2.3).

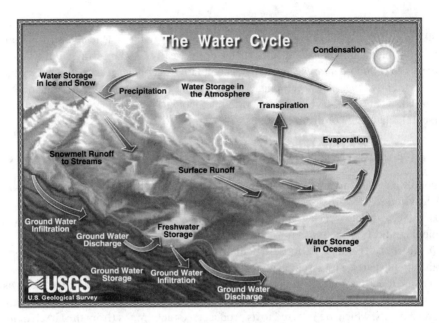

FIGURE 2.3 Diagram of the hydrologic cycle. Groundwater, surface water, and water vapor in the atmosphere are interconnected. Courtesy of the U.S. Geological Survey. (http://ga.water.usgs.gov/edu/watercycle.html)

Water is constantly on the move, and it is continually changing among its three physical states, depending on temperature: liquid water, water vapor, and solid water (ice and snow) (Figure 2.4). Liquid water in oceans, lakes, and other water bodies evaporates as it is warmed by heat energy from the sun. In addition to surface water, a second significant source of water vapor is transpiration, a process whereby water vapor is released into the atmosphere by the leaves of plants. Atmospheric water vapor is transported advectively by air currents. When it encounters lower temperatures, water vapor condenses and falls to the earth as rain. If temperatures are sufficiently low, water vapor solidifies and is deposited as snow or ice. Rain and snowmelt percolate down through the soil and into groundwater. If more rain falls than can be absorbed by the ground, the unabsorbed water and snowmelt run off into streams and lakes. The vast majority of global fresh water, some 96%, is stored underground. Only 4% of the world's freshwater is found in lakes, rivers, and other surface-water bodies. Groundwater, like surface water, flows downhill, its velocity and direction determined by the steepness of the terrain and the geologic formations it encounters. The hydrologic cycle is completed when groundwater discharges to a stream, lake, or ocean, where it once again either evaporates or is taken up by plants and transpired into the atmosphere (Figure 2.3).

Groundwater can play a significant role in transporting toxic chemicals through the environment. The location and movement of groundwater depends on geologic formations called *aquifers*. Underground aquifers discharge to surface-water bodies; they are recharged by water percolating down from the ground surface. Aquifers come in two general types: unconsolidated aquifers, in which subsurface soil or sand

Environmental Pathways of Toxic Chemicals

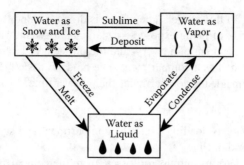

FIGURE 2.4 The physical states of water. The hydrologic cycle depends on temperature-driven transitions among the three physical states of water: solid (snow and ice), liquid, and vapor. The total mass (amount) of water on planet Earth is a constant. Global temperatures determine how much water is present as a liquid, a vapor, or a solid. Global warming shifts water from ice and snow to liquid water and water vapor.

is saturated with water; and consolidated aquifers, consisting of bedrock containing water in and among the rocks.

Aquifers vary enormously with respect to the volume of water they can hold and the speed with which water can flow through them. The volume of water in an aquifer depends on its porosity. In the case of unconsolidated aquifers, porosity is determined by the number and size of pores within and between soil particles. In bedrock aquifers, water is held, not in pores, but in fractures or holes in the rock. Spaces holding water in bedrock are referred to as secondary porosity, and the kinds of spaces depend on the geologic features of the bedrock. For example, a hard, crystalline bedrock may hold water in cracks and fractures, whereas a soft bedrock such as karst limestone may have undergone erosion by groundwater over millions of years, resulting in a "Swiss cheese" network of large fissures and holes in which water is held.

The velocity of groundwater flow depends on the permeability of an aquifer. Permeability is a function of the connectedness among water-holding spaces. The greater the connectedness among pores (in unconsolidated aquifers) and among fissures and holes (in bedrock aquifers), the greater is the permeability of the aquifer, and the faster groundwater can flow through it. Groundwater velocities range from inches per day in some unconsolidated aquifers to miles per day in some karst limestone aquifers. The rate at which an aquifer becomes contaminated by a toxic chemical depends on its permeability.

2.8 ASSESSING AND MANAGING EXPOSURE

As noted at the beginning of this chapter, limiting exposure is the preeminent strategy for managing risks from toxic chemicals. But given that myriad fate and transport mechanisms determine the paths of toxic chemicals through the environment, is it possible to make useful predictions about when and to what extent an at-risk population will be exposed?

The mass-balance model offers a conceptually simple and elegant approach to organizing available information about chemical fate and transport for purposes of assessing exposure. In this model, a boundary is drawn around the part of the environment that is at risk of being exposed—e.g., a city, a groundwater aquifer, or the fish population in a segment of a river—and the physical or biological space inside the boundary is defined as a compartment. The modeler asks, "How does the mass of a toxic chemical inside the compartment change over time?" To address this question, the modeler:

1. Identifies inputs, or loadings, into the compartment and sums them over a specified period of time
2. Identifies the various outputs, or losses, from the compartment during the same time period and sums them
3. Subtracts the outputs from the inputs to determine the net change in the mass of the chemical in the at-risk compartment over the specified period of time

In the event that the net change is zero, the masses of the chemical that enter and leave the compartment are the same, and the compartment is said to be at steady state.

While elegant in concept, the mass-balance model is not easy to apply to real-world fate and transport scenarios. A chemical typically partitions into not one, but several environmental media and also into biota, so that not one, but multiple compartments are needed to calculate mass balance. A separate set of mass-balance equations is required for each compartment, and the equations for each compartment need to be consistent with the equations for every other compartment. The need for mathematical consistency entails, in turn, the collection of extensive data on contaminant concentrations, an undertaking that is both expensive and time consuming. While collecting concentration data and developing internally consistent sets of differential equations increase the mass-balance model's fit to the "real world," such refinements also increase the model's mathematical complexity. Risk assessors have noted that the greater the mathematical complexity of a fate and transport model, the harder it is for risk managers and the general public to understand, and the less likely it is to be used to support real-world risk-management decisions.

Fortunately, precise mathematical descriptions of a toxic chemical's environmental fate and transport are rarely necessary for crafting a useful exposure assessment and an effective risk-management strategy. Rather, a basic working knowledge of how a chemical is distributed in the environment is almost always sufficient. The environmental distribution of most chemicals can be estimated by means of a qualitative application of the mass-balance concept. The first step is to identify the one environmental compartment that is of greatest concern from the standpoint of risk. Next, fate and transport processes that affect the mass of chemical entering and leaving the compartment of concern are identified and arranged according to their magnitude. When major inputs and outputs are summed in this simpler calculation, a surprisingly useful mass-balance model often emerges. The reason a simplified mass-balance model can work is that often there is one environmental compartment that is of overriding concern, and the inputs and outputs of that compartment may be dominated by just one or two fate and transport processes. Though qualitative or, at

Environmental Pathways of Toxic Chemicals

best, semiquantitative, a simplified mass-balance model can be useful from the practical standpoint of managing toxic chemical risk, because it can serve to focus limited resources on the environmental pathways that are most likely to cause harm.

The purpose of this chapter has been to introduce the broad principles underlying environmental exposure to toxic chemicals. The next chapters look at how we know that exposure to toxic chemicals produces harmful effects.

STUDY QUESTIONS

1. Give four examples of diffusive transport (partitioning) processes that distribute chemicals into water, air, and soil/sediment.
2. Calculate concentrations in water:
    a. Assume that 30 mg of benzoic acid is dissolved in 150 ml of water. The molecular weight of benzoic acid is 122 g/mol (grams per mole). What is the concentration of benzoic acid on a mass basis, i.e., mg/L (milligrams benzoic acid per liter of water)? On a mole basis, i.e., mol/L (moles benzoic acid per liter of water)?
    b. Assume, instead of benzoic acid, that 30 mg of methanol is dissolved in 150 ml of water. The molecular weight of methanol is 22 g/mol. What is the concentration of methanol on a mass basis? On a mole basis?
3. A lake in the Adirondack Park in upstate New York receives a significant amount of "acid rain" as a result of being downwind from coal-burning electricity-generating plants in the Midwest, and its sulfuric acid content has steadily increased over the years. The pH of the lake has decreased from 6.8 to 5.3. At the new pH of 5.3, what is the concentration of free H^+ ion in the lake in units of moles per liter (mol/L or molarity)?
4. A lagoon at a hazardous chemical waste site contains 200,000 U.S. gallons of sludge. The sludge has a density of 1.2 g/ml. The sludge is laced with 223 ppm of phenol on a wet-weight basis. Phenol has a molecular weight of 106 g/mol.
    a. How many liters of sludge are there in the lagoon? How many cubic meters? How many cubic feet? How many cubic yards? (Note: 1 L = 0.26 U.S. gal; 1 m³ = 1000 L; 1 m = 1.094 yd; 1 m = 3.281 ft)
    b. How many kilograms of sludge are there in the lagoon? How many pounds? How many tons? (Note: 1 kg = 2.205 lb; 1 ton = 2000 lb)
    c. What is the total mass of phenol present in the lagoon, in kilograms?
    d. What is the molar concentration of phenol in the lagoon sludge?
5. Describe how the octanol:water partition coefficient, K_{ow}, is determined experimentally. If the concentration of benzene is 3.7 ppm in water, what is its concentration in octanol, assuming standard laboratory conditions? (Hint: See Table 2.2 for log K_{ow} value for benzene.)
6. An alternative approach to estimating the relative solubility of a compound in water versus nonpolar solvents is to determine the olive-oil:water partition coefficient. The olive-oil:water partition coefficient for the pesticide DDT is 1775. If the concentration of DDT is 3 mg/L in olive oil, what is

its concentration in water with which the olive oil is in contact, assuming standard conditions?

7. The concentration of the insecticide lindane in subsurface soil at an abandoned pesticide manufacturing plant is 88 ppm. Assuming continuous contact of contaminated soil with an unconsolidated aquifer, estimate the concentration of lindane in groundwater, assuming standard conditions. (Hint: See Table 2.2 for the log K_{oc} value for lindane.)

8. Give one example of advective transport involving water, one involving soil, and one involving air.

9. The solvent trichloroethylene (TCE) is sorbed to surface soil at a concentration of 550 ppm in an open field near the motor pool of a military base, where it is used to clean and degrease engines. Winds pick up contaminated dust from the field, transport it hundreds of miles, and deposit it on a small town, among other places. An average of 0.0003 pounds of dust from the military base are dry-deposited daily on each acre of the town; the town covers a total of 2,317 acres. Give qualitative descriptions of the diffusive and advective transport processes by which TCE is conveyed from the military base to the city. Calculate the average total mass of TCE that falls on the town each year.

10. Rock salt, or $CaCl_2$, is used to keep roads and parking lots free of ice and snow in winter. In a town in upstate New York, salty water runs off into roadside ditches, which empty into a stream, which carries the salt to a lake. Local volunteers collect a water sample at a site on the stream. They determine that the cross-section of the stream at the sampling site measures 6.2 m² and the current velocity is 1.3 m/sec. They take the water sample to a lab, which determines the chloride concentration to be 144 mg/L. Name the processes of diffusive and advective transport by which rock salt moves from being solid granules on icy pavement to chloride and calcium ions in the stream. Assuming the snow melt lasts four days and the chloride concentration remains constant at the sampling site, calculate the load of chloride that is transported past the sampling point in four days.

11. Describe the hydrologic cycle. Include the following components of the cycle in your description: temperature, transitions between physical states of water, plants, advective transport, groundwater, gravity, streams, oceans.

12. Calculation of heavy metals (cadmium) content:
    a. The bioconcentration factor (BCF) for cadmium in a species of fish is 81 ($10^{1.91}$) L/kg. Calculate the concentration of cadmium in this species when the concentration of cadmium in the lake it inhabits is 0.014 mg/L. Express the cadmium concentration in fish flesh in each of the following interchangeable units: mg/kg, ppm, µg/kg, ppb, parts per thousand, and parts per hundred (percent).
    b. The average weight of the fish is 3.2 kg. What is the total mass (amount) of cadmium in the average fish? Express cadmium mass in each of the following interchangeable units: mg, µg, ounces. (Note: 1 g = 0.03527 oz)

c. The fish inhabit a small lake in the Adirondack Park with a surface area of 2.6 hectares and a mean depth of 9 m. What is the volume of the lake? What is the mass of cadmium dissolved in the lake, in kg? (Note: 1 hectare (ha) = 10,000 m²)

d. A recent census of the fish species of concern gave a count of 5,000 fish. Calculate the total mass of cadmium in this population of fish, in kilograms.

e. What is the ratio of the total mass of cadmium in the fish population to the total mass of cadmium dissolved in the lake? Is this mass ratio less than, equal to, or greater than the BCF? Explain the reason for any difference between this mass ratio and the BCF.

13. Define each of the following terms: unconsolidated aquifer, consolidated aquifer, porosity, permeability.

14. A farmer in Africa sprays the insecticide DDT on his fields for five consecutive years. His crop yields improve, while the average surface soil concentration of DDT increases from 0 to 25 mg/kg.

   a. Assuming the soil density is 1.6 kg/L, what is the total mass of DDT in one hectare of soil to a depth of 0.2 m?

   b. Assuming the bioconcentration factor for DDT in earthworms is 8 kg kg^{-1}, calculate the concentration of DDT in an earthworm.

   c. Migrant birds find the farmer's fertile fields an inviting source of earthworms. Each bird eats an average of 12 worms a day. Assuming worms weigh an average of 2 g, and an average bird weighs 670 g, calculate a bird's daily intake of DDT, in units of milligrams DDT per kilogram body weight (mg/kg/day).

ANSWERS TO STUDY QUESTIONS

1. Evaporation, volatilization, dissolution, adsorption.
2. a. 200 mg/L benzoic acid; 1.64×10^{-3} mol/L benzoic acid.
   b. 200 mg/L methanol; 9.1×10^{-3} mol/L methanol.
3. $[H^+] = 10^{-5.3} = 5.01 \times 10^{-6}$ mol/L.
4. a. 769,231 liters; or 769 cubic meters; or 27,186 cubic feet; or 1,007 cubic yards of sludge.
   b. 923,077 kilograms; or 2,035,385 pounds; or 1,018 tons of sludge.
   c. 205.8 kilograms of phenol.
   d. 2.5×10^{-3} mol/L of phenol.
5. See text for determination of octanol:water partition coefficient (K_{ow}). Benzene: $\log K_{ow} = 2.13$, therefore $K_{ow} = 10^{2.13} = 134.9$.
   K_{ow} is the ratio of the concentration of benzene in octanol to its concentration in water.
   Benzene concentration in octanol = 3.7 ppm × 134.9 = 499 ppm.
6. Partition coefficient indicates DDT is 1,775 times more concentrated in olive oil than in water when the two liquids are in contact.
   DDT concentration in water = (3 mg/L)/1,775 = 0.00169 mg/L.

7. Lindane: Logarithm of soil:water partition coefficient (log K_{oc}) = 2.86, therefore $K_{oc} = 10^{2.86} = 724.4$ L/kg.
   Concentration of lindane in soil = 88 ppm = 88 mg/kg.
   Concentration of lindane in groundwater = (88 mg/kg)/(724.4 L/kg) = 0.12 mg/L.
8. Advective transport by water: Transport of dissolved chemical or suspended soil or sediment particles by flowing stream. Advective transport by soil: Erosion of soil particles containing sorbed chemical; sedimentation of soil particle with sorbed chemical to bottom of stream, lake, or ocean. Advective transport by air: Transport by air currents of chemicals as gases mixed with air, of chemicals sorbed to dust particles, or of chemicals dissolved in airborne water droplets.
9. Diffusive transport: TCE sorbs to soil particles. Advective transport: Wind transports TCE-containing dust particles; dust particles containing TCE fall to ground.
   Yearly TCE mass = (0.0003 pounds/acre/day) × (2,317 acres) × (365 days/year) = 254 pounds/year.
10. Diffusive transport: Salt dissolves in water. Advective transport: Water runs off road, transporting dissolved salt to ditch and stream.
    Flow (volume of water per unit time) = 1.3 m sec^{-1} × 6.2 m^2 = 8.1 m^3 sec^{-1}.
    Chloride concentration = 144 mg L^{-1} = 144 g m^{-3}.
    Chloride load in 4 days = 144 g m^{-3} × 8.1 m^3 sec^{-1} × 86,400 sec day^{-1} × 4 days = 4.03 × 10^5 kg.
11. See Section 2.7.
12. a. Cadmium concentration in fish = 0.014 mg L^{-1} × 81 L kg^{-1} = 1.13 mg kg^{-1}, or 1.13 ppm, or 1,130 µg kg^{-1}, or 1,130 ppb, or 0.00113 parts per thousand, or 0.000113 percent.
    b. Cadmium mass in average fish = 1.13 mg kg^{-1} × 3.2 kg fish^{-1} = 3.63 mg fish^{-1}, or 3,630 µg fish^{-1}, or 0.000128 oz fish^{-1}.
    c. Volume of lake = (2.6 ha) × (10^4 m^2 ha^{-1}) × (9 m) = 234,000 m^3.
       Mass of dissolved cadmium = (0.014 mg L^{-1}) × (234,000 m^3) × (1,000 L m^{-3}) = 3.276 × 10^6 mg = 3.276 kg.
    d. Total cadmium load in fish population = (3.63 mg fish^{-1}) × (5,000 fish) × (10^{-6} kg mg^{-1}) = 0.018 kg.
    e. Mass ratio cadmium$_{fish}$/cadmium$_{lake}$ = 0.018 kg/3.276 kg = 0.005.
       BCF = 81 L kg^{-1}.
    Bioconcentration is driven at the molecular level by the cadmium concentration in water, not by the total mass of cadmium in the lake. The bioconcentration factor is a ratio of concentrations, not a ratio of masses.
13. See Section 2.7.
14. a. Soil volume = (1 ha) × (10^4 m^2 ha^{-1}) × (0.2 m) × (10^3 L m^{-3}) = 2 × 10^6 L.
       Soil mass = (2 × 10^6 L) × (1.6 kg L^{-1}) = 3.2 × 10^6 kg.
       DDT mass = (25 mg kg^{-1}) × (3.2 × 10^6 kg) × (10^{-6} kg mg^{-1}) = 80 kg.
    b. DDT concentration in earthworm = (25 mg kg^{-1}) × (8 kg kg^{-1}) = 200 mg kg^{-1} or 200 µg g^{-1}.

c. One bird's daily DDT intake = 12 worms day^{-1} × 2 g worm^{-1} × 0.2 mg DDT g^{-1} = 4.8 mg DDT day^{-1}.
One bird's daily DDT intake on a mass basis = (4.8 mg DDT bird^{-1} day^{-1})/0.67 kg bird^{-1} = 7.16 mg DDT kg^{-1} day^{-1}.

REFERENCES

Eisler, R. 2000. *Handbook of chemical risk assessment.* Vol. I, *Metals.* Boca Raton, FL: Lewis Publishers.
Jorgensen, E. P., ed. (Sierra Club Legal Defense Fund). 1989. *The poisoned well: New strategies for groundwater protection.* Washington, DC: Island Press.
Miller, G. T. Jr. 1991. *Environmental science.* 3rd ed. Belmont, CA: Wadsworth Publishing.
Montgomery, J. H. 2000. *Groundwater chemicals desk reference.* 3rd ed. Boca Raton, FL: CRC Press.

SUGGESTED READING

Brusseau, M. L. and H. L. Bohn. 1996. Chemical processes affecting contaminant fate and transport in soil and water. In *Pollution science,* ed. I. L. Pepper, C. P. Gerber, and M. L. Brusseau, 63–75. San Diego: Academic Press.
Lyman, W. J. 1995. Transport and transformation processes. In *Fundamentals of aquatic toxicology,* 2nd ed., ed. G. M. Rand, 449–492. Washington, DC: Taylor & Francis.
Mackay, D. and S. Paterson. 1993. Mathematical models of transport and fate. In *Ecological risk assessment,* ed. G. Suter, 129–152. Chelsea, MI: Lewis Publishers.

3 Dose–Effect
The Foundation of Toxicological Science

3.1 INTRODUCTION

Common sense tells us that coming in contact with a toxic chemical—touching it, inhaling it, ingesting it—makes us sick. The more contact, the sicker people are likely to get. For example, if a student spills acetone in his organic chemistry class, or if a homeowner spills paint thinner in her garage, those nearest the spill inhale the most fumes, and they are more likely to feel nauseous or dizzy than those at a distance who inhale less fumes. At the same time, some people who are farther away and inhale less fumes may be particularly sensitive and end up feeling just as sick as those nearest the spill.

The key to managing toxic chemicals judiciously is the dose–effect relationship. This relationship correlates degrees of exposure with manifestations of toxicity. It shows the percent of exposed individuals who get sick as a function of their degree of contact with a toxic chemical. The dose–effect relationship has the potential to serve as a toxicological road map for how to use chemicals to improve the quality of life while lowering to acceptable levels the risk of making people sick.

3.2 ETHICAL DILEMMAS AND THE PROTECTION OF PUBLIC HEALTH

In order to investigate the dose–effect relationship for a toxic chemical, it is necessary to conduct experiments in which groups of individuals are exposed to different doses. Herein lies a dilemma: How can the health effects of chemicals be investigated when it is unethical to test chemicals on humans? Toxicologists have addressed this dilemma in various ways, most directly and successfully by testing chemicals on laboratory animals as surrogates for human beings. Small mammals such as rats, mice, and guinea pigs are close evolutionary cousins of *homo sapiens*, and their physiological responses to toxic chemicals are similar, although not identical, to our own. The close physiological similarities mean that small mammals are reasonable models for investigating the effects of chemicals on humans.

Animal testing, of course, presents its own set of ethical dilemmas. In an effort to address concerns about the suffering of animals in laboratory tests, federal guidelines mandate that test animals be cared for in an ethical manner. Occasionally, animal testing may be avoided by using strategies such as structure-activity relationships or *in vitro* ("in glass," or test tube) laboratory tests to identify toxicity in a chemical (Chapter 5). However, given the current state of knowledge, it is simply not possible to simulate the full range and complexity of the interactions that take place between a toxic chemical and a living organism. From the standpoint of assessing chemical toxicity, it can be stated unequivocally that there is no viable substitute for animal testing when the goal is to predict the effects of chemical exposure on human health with a reasonable degree of confidence. The best that can be done is to treat test animals ethically and minimize the numbers that are used.

Animal toxicity testing is described in detail in Chapter 5. The results of animal toxicity testing form the basis of human health risk assessments (Chapter 8). In addition to human health, there is also growing concern with the effects of toxic chemicals on fish, wildlife, and plants and on the ecosystems that they—and ultimately we as humans—depend upon for our lives and livelihoods. The dose–effect relationship is investigated directly in several species that can be maintained in the laboratory, for example, rainbow trout, fathead minnows, and ring-necked pheasant. Results may be extrapolated to other species and used to support ecological risk assessments (Chapter 9).

3.3 PRELIMINARY INVESTIGATIONS OF TOXICITY

Biological systems span many levels of organization: cells, tissues, individual organisms, populations of organisms, communities of species, and ecosystems that encompass multiple biotic (living) communities as well as abiotic (nonliving) components. Chemical toxicity may be investigated at any level of biological organization; however, the two levels investigated most frequently are individual organisms and populations of organisms. The quantitative assessment of toxic chemical risk is based on studies of populations. Preliminary investigations of toxicity are often conducted on individual organisms. Studies on individual organisms fulfill two important functions with respect to risk assessment: First, they provide valuable information with which to design effective population-level studies, such as the types of toxicity a chemical produces and the dose range associated with each toxic effect. Second, studies in individual organisms can provide evidence of a chemical's mechanism of toxicity. While studies in individual organisms are an important part of a toxicologist's toolkit, quantitative risk assessment and predictions of chemical disease are based on studies of how populations, not individuals, respond to a chemical. Risk assessment is a statistical tool that is designed to help manage toxic chemical risk to populations as a whole (Chapters 8 and 9). It is not feasible to tailor risk management to individual organisms within a population.

3.4 THE QUANTAL DOSE–EFFECT RELATIONSHIP: THE WORKHORSE OF RISK ASSESSMENT

In population-level studies, attention is focused on the relationship between exposure to a chemical and the incidence of toxicity in the test population. The size of the test population depends on the statistical power required of the test (see Chapter 4) balanced against the cost of maintaining large numbers of test animals in the laboratory. In a population-level study, test animals are divided into several dosage groups and one control group. Each group is equal in size to the test population. The same dose of chemical is administered to all the animals in a particular dosage group. A different dose is administered to each dosage group, while the control group receives no chemical at all. A window of time is specified, and the animals are observed carefully for signs of the specific toxic effect, or endpoint, that is under investigation, for example, cancer, blindness, a birth defect, or death. The number of animals in each dosage group that manifest the endpoint of concern within the specified window of time is counted. A hypothetical data set based on a test population of 30 animals is presented in Table 3.1. The numbers of stricken animals in the dosage groups are the raw data from which the incidence of toxicity in the test population is determined as a function of dose. The data in Table 3.1 are graphed in Figure 3.1a and b.

Raw data on the incidence of toxicity may be analyzed in either of two ways. First, incidence may be analyzed as the additional, or incremental, number or percent of animals that are stricken at successively higher doses of the test chemical (Figure 3.1a). Second, analysis of the exposed population may focus on the total, or cumulative, number or percent of animals that are affected as the dose is raised (Figure 3.1b).

TABLE 3.1
The Quantal Dose-Effect Relationship: Frequency of Anemia as a Function of Chemical Dose in a Test Population of 30 Rats

Dosage Group [a] (mg/kg)[b,c]	Incremental No. Rats Manifesting Anemia[d]	Cumulative No. Rats Manifesting Anemia[e]	Incremental No. as Percent[f]	Cumulative No. as Percent[f]
0	0	0	0	0
5	1	1	3.3	3.3
10	2	3	6.7	10.0
20	4	7	13.3	23.3
50	5	12	16.7	40.0
100	7	19	23.3	63.3
200	6	25	20.0	83.3
500	3	28	10.0	93.3
1000	2	30	6.7	100.0

Note: See Figures 3.1a and b for graphs of data in this table.
[a] Each dosage group contains 30 animals, as does the control (zero dose) group.
[b] Units of dose are milligrams of chemical per kilogram body weight of the test animal.
[c] Dosage axes (x-axes) of graphs in Figure 3.1 are scaled in logarithmic units.
[d] Number of new anemia cases in each successively higher dosage group.
[e] Total number of animals with anemia at each successively higher dosage group.
[f] Percent calculated by dividing 30 into either the incremental number or the cumulative number of rats manifesting anemia and multiplying by 100.

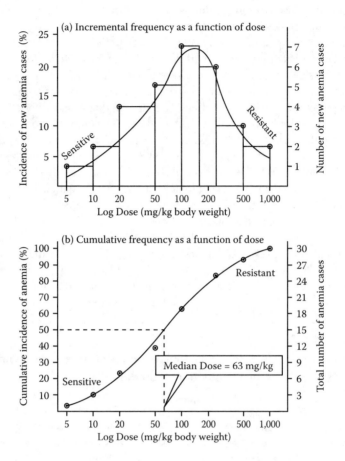

FIGURE 3.1 The quantal dose-effect relationship: Frequency of anemia as a function of chemical dose. (a) Incremental frequency as a function of dose. The curve, called a frequency histogram, plots the incremental frequency data from Table 3.1. Anemia is the quantal effect, or endpoint, and is defined in this hypothetical study as a decrease of 35% or more in the volume of red blood cells. Test animals with less than a 35% decrease are considered not to be anemic. Animals that manifest anemia at low doses (left end of frequency distribution) are termed "sensitive," while animals that do not manifest anemia unless exposed to high doses (right end of frequency distribution) are termed "resistant." (Note: The x-axis is scaled in logarithmic units, while the y-axis is scaled in linear units. A graph with one linear axis and one logarithmic axis is referred to as semilogarithmic.) (b) Cumulative frequency as a function of dose. The curve, which is sometimes referred to as a quantal dose-effect curve, plots the cumulative frequency data from Table 3.1. The cumulative frequency curve is sigmoidal (S-shaped). The median dose may be determined graphically from the inflection point of the sigmoidal curve. The quantal dose-effect curve offers practical advantages as an analytical tool and is widely used in toxicology.

Dose–Effect

3.4.1 Analysis of Incremental Toxicity

When the incremental number or percent of stricken test animals in each successive dosage group is plotted against the logarithms of the chemical doses, the result is a frequency distribution or histogram (Figure 3.1a). A histogram shows the frequency with which new cases of toxicity develop at each successively higher dose of the test chemical. For example, in the dosage groups receiving 20 and 50 mg of test chemical per kilogram body weight, the total number of affected animals is 7 and 12, respectively; therefore, the incremental number of affected animals in the 50-mg/kg dosage group is 5 (12 − 7 = 5) (Table 3.1).

The frequency histogram in Figure 3.1a approximates a bell-shaped curve, a common result in chemical toxicity testing. Frequency distributions that are bell-shaped are referred to by statisticians as being normal, or Gaussian, distributions. A bell curve demonstrates that susceptibility to chemical toxicity varies among individuals. It demonstrates further that this individual variability is randomly distributed in a population: Some individuals are more susceptible than others. However, it is essentially impossible to predict which individuals in a population are more susceptible and which are less susceptible. Susceptibility to toxicity is part of the more general phenomenon of biological variability among individual members of a species. Many factors contribute to biological variability as it relates to toxicity, for example, age or stage in the life cycle (e.g., adult, newborn, etc.), nutritional status, preexisting disease, and genetic makeup including gender. Even in a population of closely matched individuals, such as inbred strains of laboratory rats of the same sex and similar age, differences in susceptibility to toxicity, while smaller than those found in wild populations, are still observed. The shape of the frequency histogram indicates the degree of variability among individuals in the test population: The broader the histogram, the greater the variability, and the narrower the histogram, the more closely individuals in the test population tend to resemble one another with respect to their susceptibility to the specified toxic effect of the test chemical.

While the breadth of the histogram provides a general indication of the variability of the biological response of the test population to a test chemical, the numbers on the x-axis indicate the actual doses of a chemical that produce the toxic effect of concern. One of these numbers, the midpoint, is commonly used to characterize the test population as a whole. The midpoint of any frequency distribution is a statistic called the median. Because this is a dose–effect histogram, the median is referred to as the median dose.

The definition of the median dose is that half of the animals in the test population manifest the specific toxic endpoint (e.g., cancer, blindness, anemia, birth defect, death) at doses below the median, and the other half manifests the specific toxic effect at doses above the median. In other words, the median is the midpoint of the frequency distribution of the toxic effect of concern. It is important to note that the median dose is not the same as the mean, or average, dose. The average is calculated as the sum of all the doses of the test chemical divided by the number of dosage groups. It turns out that the mean dose coincides with the median dose only if the dose–effect histogram is a perfect bell-shaped curve. In practice, the frequency distribution is seldom perfectly bell-shaped, and the mean is rarely equal to the median.

For example, the average dose in Table 3.1 is 236 mg/kg, while the median dose is 63 mg/kg (Figure 3.1b). The distinction between mean and median dose is important because the mean and median provide related but different kinds of information about chemical toxicity: The median dose indicates half-maximal toxic effect frequency; the mean is the average dose the population is exposed to.

3.4.2 Analysis of Cumulative Toxicity

Histograms such as that in Figure 3.1a contain all of the raw information that is needed to analyze a specific toxic effect of a chemical on a test population. However, as an analytical tool, a histogram is awkward and imprecise. For example, it is not easy to determine the median dose because most histograms are not perfectly bell-shaped, and therefore the median dose does not fall exactly in the middle. For increased analytical precision and ease, a cumulative frequency curve, which is also referred to as a quantal dose–effect curve, is used instead of the frequency-response histogram. This analytical tool focuses on the cumulative, or total, number of stricken animals at each dose. In effect, it adds up the y-values of the histogram and plots the cumulative sum of the y-values corresponding to each dose. The result is a sigmoidal, or S-shaped, curve (Figure 3.1b). The inflection point of the graph (where the curvature changes direction) corresponds to the midpoint of the frequency distribution, the point at which the number of new or incremental cases of stricken test animals stops increasing and starts decreasing. In other words, the inflection point corresponds to the median dose. The significance of the median dose, again, is that the doses that produce the specified toxic effect in the first 50% of the test animals are all less than the median dose, while the doses that produce the specified toxic effect in the second 50% of the test animals are greater than the median dose. The median dose is determined by drawing a horizontal line from the 50% value on the y-axis to the inflection point of the sigmoidal cumulative dose–effect curve, then dropping a perpendicular line from the inflection point to the x-axis. The point of intersection with the x-axis is the median dose (Figure 3.1b).

The median dose is often used as a shorthand characterization of the entire dose–effect relationship between a chemical and a test population. Generally, the median dose for a toxic effect is referred to as the 50% toxic dose, or TD_{50}. When the toxic endpoint is death, the median dose is referred to as the 50% lethal dose, or LD_{50}. While the median dose is a useful shorthand characterization for the majority of a test population, it has an unfortunate tendency to obscure members of the population that manifest toxicity at very low doses, referred to as sensitive, as well as members of the population that are not affected unless exposed to very high doses, referred to as resistant (Figure 3.1). Sensitive individuals are of particular concern because they are the most vulnerable members of a population to chemical assault. These individuals may make up, very roughly, 1% to 5% of a population, depending on how the dose threshold for sensitivity is defined. Doses that impact sensitive individuals need to be considered along with the median dose when assessing chemical risk.

The cumulative dose–effect curve (Figure 3.1b) and the frequency-response histogram (Figure 3.1a) offer two distinct approaches to analyzing data from toxicity tests. Nevertheless, the two graphs are fully interconvertible. The cumulative dose–effect curve may be constructed by summing over the frequency histogram. Conversely, the frequency histogram may be constructed from the cumulative dose–effect curve by calculating incremental response frequencies at successive doses. However, the cumulative dose–effect curve offers practical advantages over the frequency-response histogram. One advantage is the increased accuracy of determining the median dose from the inflection point. Another is that cumulative quantal dose–effect curves for several different populations or several different chemicals are readily displayed together in the same graph, facilitating comparisons among different populations or different chemicals.

The cumulative dose–effect curve is used extensively to characterize chemical toxicity. It provides a foundation for all forms of chemical risk assessment, including human health risk assessment and ecological risk assessment. Because this relationship will be encountered again and again when evaluating risk, several important features of the cumulative dose–effect curve are worth emphasizing:

1. It applies to populations, not individuals.
2. It concerns the frequency of occurrence of a single toxic effect of a specified magnitude, referred to as a quantal effect, or endpoint. Examples are death, a specific birth defect, infertility, cancer, liver disease, anemia, kidney failure, or a cardiac arrhythmia. It does not concern different magnitudes of the same toxic effect nor does it involve a spectrum of toxic effects.
3. It is a statistical relationship that describes the incidence of a specified toxic effect as a function of the dose that a test population is exposed to.
4. A key feature of a quantal dose–effect relationship is a statistic: the median dose, variously annotated as the TD_{50} or LD_{50}, which marks the middle of the exposed population with respect to its susceptibility to the quantal toxic effect. Half the population manifests the specified toxic effect at doses less than the TD_{50}, and half the population manifests the specified toxic effect at doses greater than the TD_{50}.
5. The median dose is distinct from the mean dose. The mean dose is the average dose to which a population is exposed. In practice, the mean dose rarely coincides with the median dose.
6. The slope is a general indicator of biological variability with respect to the susceptibility of individuals in the test population to the specified toxic effect of the test chemical. The shallower the slope, the greater is the biological variability among individuals in the exposed population with respect to their vulnerability to the specified toxic effect.

3.5 THE GRADED DOSE–EFFECT RELATIONSHIP

A toxicity investigation in a population of animals provides a statistical tool for making decisions about how to manage a risk arising from exposure to a toxic chemical. A toxicity investigation in an individual animal or a small number of animals lacks statistical power, and therefore it is considered less useful for risk-management purposes. However, it provides important preliminary data on the toxic effects of a chemical and the dose ranges over which the effects occur. Such information is needed to design population-level toxicity studies; it is also important information in its own right.

Besides providing preliminary data for population-level toxicity studies and chemical risk assessment, small numbers of animals may also be used for detailed investigations of the mechanism of chemical toxicity. Studies of mechanism generally involve investigations at several levels of biological organization, such as the whole organism, the affected tissue, the impacted cells, and ultimately the biological molecules that are damaged by toxicant molecules. Typically, the investigation of the mechanism of toxicity begins with studies of the individual organism and eventually ends with studies at the molecular level. Because of the complexity of molecular events and the time and expense involved in performing research to elucidate them, mechanisms of toxic action are known at the molecular level for only a limited number of chemicals.

A study in a single, intact animal involves a determination of the magnitude, or intensity, of a toxic effect as a function of the dose of the test chemical administered to the individual animal. Data are presented in Table 3.2 from a hypothetical investigation of anemia in an individual rat that belongs to the same general population as the rats in the population study presented in Table 3.1 and Figure 3.1. The magnitude of the toxic effect (decrease in red blood cells) is determined for each dose of chemical that is administered to the individual test animal, and the data are analyzed by plotting the magnitude of the effect on the y-axis against the dose on the x-axis. The resulting graph is called a graded dose–effect curve.

Often, dose is scaled in linear units, resulting in a curve that rises steeply at low concentrations and approaches a plateau as the dose is increased (Figure 3.2a).

TABLE 3.2
The Graded Dose-Effect Relationship: Magnitude of Anemia as a Function of Chemical Dose in an Individual Rat

Dose (mg/kg body weight)	Logarithm of Dose[a] (mg/kg body weight)	Magnitude of Anemia[b] (% decrease in red blood cells)
Control	Control	None
0.5	−0.30	8.8
1	0.00	15.6
2	0.30	25.4
3	0.48	32.3
5	0.70	41.2
10	1.00	51.9
20	1.30	59.6
50	1.7	65.4

Note: See Figures 3.2a and b for graphs of data in this table.

[a] Logarithm to the base 10. Thus, $0.5 = 10^{-0.3}$, $3 = 10^{0.48}$, $10 = 10^{1.00}$, etc.

[b] The magnitude of anemia is determined on the basis of a decrease in red blood cells. The maximum magnitude of anemia is defined as a 70% decrease in red blood cells in this hypothetical study.

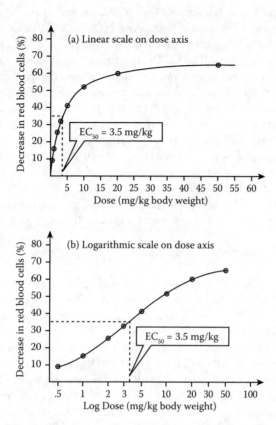

FIGURE 3.2 The graded dose-effect relationship: Magnitude of anemia as a function of dose in an individual rat. (a) Analysis using a linear scale on dose axis. Data from Table 3.2 are plotted on this graph and show increases in the magnitude of anemia as a function of increasing chemical dose. The dose axis is scaled in linear units, resulting in a curve that rises steeply at low doses and approaches a plateau at higher doses. Decreases in red blood cells are used as an index of anemia, and the maximum magnitude of anemia is defined as a 70% decrease in red blood cells. The dose that produces a half-maximal effect (i.e., a 35% decrease in red blood cells) is defined as the 50% effective concentration, or EC_{50}. The EC_{50} is determined to be 3.5 mg/kg by drawing a horizontal line from the point on the y-axis corresponding to the half-maximal effect magnitude and dropping a perpendicular line from the point of intersection with the curve to the x-axis. Note that this particular animal's EC_{50} is significantly lower than the median dose of 63 mg/kg (Figure 3.1b), indicating it may be characterized as "sensitive" compared with other individuals in the exposed population. (b) Analysis using a logarithmic scale on dose axis. This is the same curve as in Figure 3.2a except that the x-axis has been changed from a linear scale to a logarithmic scale. Linear distances on the x-axis now correspond to exponents of 10. Thus, $2 = 10^{0.3}$, so that 2 is located 0.3 of the distance between 1 and 10; $5 = 10^{0.7}$, so that 5 is located 0.7 of the distance between 1 and 10, and so on. The result is a transformation of the shape of the graded dose-effect curve from a rectangular hyperbola to a sigmoidal curve. The EC_{50} corresponds to the x-coordinate of the inflection point of the sigmoidal curve.

When dose on the x-axis is scaled in logarithmic instead of linear units, the curve is sigmoidal (Figure 3.2b). The 50% effective concentration, or EC_{50}, is defined as the dose at which the magnitude, or intensity, of the toxic effect in the individual test animal is half-maximal. The EC_{50} is determined in a manner analogous to the determination of the median dose in a population study (Figure 3.1b): A horizontal line is drawn from the point on the y-axis corresponding to 50% of the maximal intensity of the harmful effect, and where the horizontal line intersects the graded dose-response curve, a perpendicular line is drawn to the x-axis. In the example illustrated in Figures 3.2a and b, the EC_{50} is found to be 3.5 mg/kg. This particular animal's EC_{50} thus places it at the sensitive end of the dose–effect relationship for the population to which it belongs (compare this animal's EC_{50} with the population dose–effect curve in Figure 3.1b).

When an animal's EC_{50} is known, the effect magnitude (intensity) in that particular animal may be calculated at any dose using the following equation:

$$\text{Effect magnitude} = \text{Maximal effect} [\text{Dose}/(\text{Dose} + EC_{50})] \quad (3.1)$$

The form of Equation (3.1) is known as a rectangular hyperbola. While Equation (3.1) describes the intensity of a physiological effect as a function of the dose of a chemical administered to a whole animal, the same formal mathematical relationship also describes interactions at the molecular level in which small molecules, such as toxic chemicals, bind to large biological molecules, such as proteins and DNA. When Equation (3.1) is used to describe harmful molecular interactions, dose becomes the concentration of small toxicant molecules, and effect magnitude is the percent of large biological molecules that are occupied by toxicant molecules. Equation (3.1), as applied to harmful molecular interactions, states that the larger the concentration of a small toxicant molecule, the greater is the proportion of biological target molecules that are occupied by it. Another name for this effect is mass action, meaning that as the concentration of the small molecule increases, the concentration of the small molecule/large molecule complex also increases. This is one example of a fundamental principle in chemistry: Increasing the concentration of a reagent in a chemical reaction results in an increase in the concentration of the product of the chemical reaction. This principle is called the law of mass action.

Most known molecular mechanisms of toxicity involve an interaction between a small toxicant molecule and a large biological molecule that fits the mass-action model and is described by a rectangular hyperbola. It is for this reason that a graded dose–effect curve for an intact organism, such as that presented in Figure 3.2, is inferred to reflect the occupation of biological molecules by toxicant molecules inside the animal's body. Stated differently, the increasing occupation of vulnerable biological molecules by toxicant molecules is the basis for the increasing intensity of the toxic effect on the whole animal as exposure increases. Thus, the law of mass action is the vehicle of toxic chemical risk—and the molecular backstory behind Paracelsus's famous dictum, "the dose makes the poison."

The distinctive features of the graded dose–effect curve can be summarized as follows:

1. It applies to individuals, not populations. It is a direct relationship between the dose of a chemical and the magnitude of its harmful effect in an individual organism. It is not a statistical relationship.
 2. It concerns the continuously increasing magnitude of one single harmful effect as a function of administering increasing doses of a chemical to an individual animal.
 3. The basis of the increasing magnitude of the harmful effect with dose is the occupation of an increasing percentage of the target population of susceptible biological molecules in the body by the molecules of the toxic chemical.
 4. The graded dose–effect curve is used to determine the dose that produces 50% of the maximal magnitude of the harmful effect in the test animal. This dose is referred to as the 50% effective dose (ED_{50}); alternatively, it is referred to as the 50% effective concentration (EC_{50}) to distinguish it clearly from the median dose in population studies. The EC_{50} refers to the dose that is administered to the whole animal, not to the concentration of the toxic chemical inside the animal's body. While the EC_{50} administered to the whole animal is not the same as the animal's internal concentration of the toxic chemical, the EC_{50} is thought to be proportional to the toxic chemical's internal concentration.

The graded dose–effect curve is used as a point of departure for designing population-level toxicity tests because it provides information about the magnitude of the harmful effect (or effects) produced by increasing doses of a chemical. This information is used to pick one particular magnitude of a harmful effect as the endpoint for a population-level study. For example, a 35% decrease in red blood cells was defined as the specific "anemia" endpoint in the hypothetical population study presented in Figure 3.1. If a decrease of 30% or 25% had been used as an endpoint instead, more animals would have become "anemic" at lower doses in the population study than did when the threshold of anemia was defined as a 35% decrease. In other words, how the toxicity endpoint is defined has a profound effect on the outcome of a population study, including the determination of the median dose.

We have seen that a fundamental principle of chemistry, the law of mass action, underlies the graded dose–effect curve (Figures 3.2a and b) and provides a molecular explanation for Paracelsus's dictum, "the dose makes the poison." In addition to toxic effects in individual organisms, the law of mass action also explains the quantal dose–effect curve that describes population-level effects (Figure 3.1b). The reasoning is as follows: In each dosage group in a population study, animals are given the same dose of a chemical. However, individual animals process and respond to the chemical differently (Chapters 6 and 7), with the result that the concentration of toxicant-target molecule complexes as well as the physiological effect of a given concentration are different for each animal. At lower doses of toxic chemical, some animals have a concentration of toxicant-target molecule complexes in their tissues that leads to a 35% or greater decrease in red blood cells while others do not. As the dose of toxicant is raised, and ever larger numbers of vulnerable biological molecules are occupied by toxicant molecules, individual differences in animals' responses are increasingly obscured. When the dose is sufficiently high, toxicant molecules occupy the fraction of target molecules needed to produce the defined endpoint of "anemia" in all the animals in the study population.

The cumulative quantal dose–effect curve in Figure 3.1b, like the graded dose–effect curve in Figure 3.2, has the mathematical form of a rectangular hyperbola:

$$\text{Effect frequency} = \text{Maximal frequency} [\text{Dose}/(\text{Dose} + \text{Median dose})] \quad (3.2)$$

The quantal dose–effect curve describes the response of a population to increasing doses of a toxic chemical. The graded dose–effect curve describes the response of an individual to increasing doses of a toxic chemical. The basis of their mathematical similarity appears to be that toxic effects at the level of the individual and at the level of the population are both driven by the law of mass action at the molecular level. As noted previously, in cases where target molecules have been identified, the extent of their occupation by toxicant molecules *in vitro* is also described by a rectangular hyperbola. The evidence is strong, although ultimately circumstantial, that mass action at the molecular level translates into dose–effect relationships at the level of the individual and the population. If there can be said to be a central dogma in toxicology, it is that the cornerstone of toxicological science, the dose–effect relationship, rests squarely on the chemical properties of toxicant and biological molecules and how they interact inside the body.

STUDY QUESTIONS

1. True or false: "A quantal dose–effect relationship quantitates the toxicity of a chemical to an individual animal as a function of dose." Explain.
2. What is the purpose of specifying an "endpoint" in a population-level study of chemical toxicity? Give four examples of endpoints.
3. Explain the difference between the "incremental incidence" of toxicity and the "cumulative incidence" of toxicity in a population-level (statistical) study of toxicity. Does either of these concepts apply to toxicity investigations in individual animals? Why or why not?
4. How is the EC_{50} defined in a toxicity study in an individual animal?
5. How is the median dose defined in a population-level toxicity study?
6. True or false: "The slope of the cumulative dose–effect curve provides a general indication of the biological variability of the test population." Explain.
7. A manufacturer is developing a new rat poison and conducts a lethality test on a population of rats drawn from an inbred strain of laboratory animals. If the lethality test were repeated on rats captured in a shipyard, would you expect the cumulative dose–effect curve to be shallower or steeper than the curve for the inbred rats?
8. How is the mean dose in a population-level toxicity study defined? Give two practical reasons why the mean dose may differ from the median dose in a "real world" data set.
9. Data showing the decrease in red blood cells as a function of test chemical dose in an individual animal are tabulated in Table 3.2 and graphed in

Figures 3.2a and b. Determine the dose of test chemical that causes a 50% decrease in red blood cells using each of the following three methods:
   a. The graph in Figure 3.2a
   b. The graph in Figure 3.2b
   c. Equation (3.1)
10. Data tabulated in Table 3.1 and graphed in Figure 3.1b show the cumulative, or total, incidence of anemia in a test population of animals as a function of chemical dose. What is the incidence of anemia at a test chemical dose of 30 mg/kg body weight? Solve this problem using:
   a. Equation (3.2)
   b. The graph in Figure 3.1b

The two answers should agree closely. Remember: The scale on the x-axis in Figure 3.1b is logarithmic, not linear. First determine the fraction of the distance between 10 and 100 that corresponds to the logarithm of 30, then find the incidence of anemia that corresponds to that distance fraction on the x-axis.

ANSWERS TO QUESTIONS AND PROBLEMS

1. False. The quantal dose–effect relationship determines the frequency of one specified toxic effect in a population that is exposed to increasing doses of a single chemical. The term *quantal* refers to the fact that the toxic effect is of a specific, known magnitude. The frequency of the manifestation of the specified toxic effect in the exposed population, i.e., the number of individuals who manifest the specified toxic effect, increases as a function of dose. However, it is impossible to predict which individuals in the at-risk population will manifest the toxic effect at any given dose.
2. An "endpoint" is the same as a "specified toxic effect." Without a known, specific, quantifiable endpoint, a statistically valid determination of the frequency of toxicity would not be possible. A few examples of specific endpoints in population-level studies of toxicity are death, infertility, a birth defect, cancer, deafness, anemia, liver disease, and IQ.
3. "Incremental incidence" refers to the number of new cases of individuals that manifest the specified toxic effect, or endpoint, at a given dose but did not manifest it at the previous, lower dose. "Cumulative incidence" refers to the sum of the individuals that manifest the specific toxic effect, from the lowest dose up to and including the current dose. Neither of these concepts applies to toxicity investigations in individual animals. Both are statistical concepts, and statistical concepts apply only to populations, not to individuals.
4. The EC_{50} is defined as the dose of the chemical administered to an individual animal (not the internal concentration of the chemical) at which the magnitude or intensity of the toxic effect is equal to half of its maximal magnitude or intensity in that individual animal. For example, if the toxic effect is deafness, then normal hearing might be defined as being able to hear sounds of two decibels intensity; deafness as being able to hear sounds

only when they are, say, 150 decibels or more; and half-deafness as half the increase in decibel level, or $(150 - 2)/2 = 74$ decibels. The EC_{50} would then be determined as the dose that produces half-deafness. Note that some toxic effects increase continuously as a function of dose (within the physiological limits of the organism, of course). Examples in humans are deafness as an adverse effect of taking aminoglycoside antibiotics, decreased breathing in response to morphine administered to alleviate pain, and anemia as a result of exposure to lead. Some toxic effects manifest not as a physiological continuum but as discrete events. Examples are death, infertility, and cancer. It is possible to determine an EC_{50} value in a test animal for a toxic effect that varies in magnitude as a function of dose but not for a toxic effect that manifests as a single discrete event.

5. The median dose is defined as the dose below which 50% of the study population manifests the specified endpoint and above which the other 50% of the study population manifests the endpoint. The median dose is not an actual dose. It is a statistic that is extrapolated from the results of a population study. It is derived from the frequency, or incidence, of manifestation of the specified toxic endpoint as a function of dose.

6. True. A shallow slope means that the manifestation of the specified endpoint of toxicity is spread out over a large dose range. The larger the range of doses required to trigger manifestation of the endpoint, the greater is the biological variability of the test population with respect to that endpoint.

7. Shallower. The genetic differences within a population of outbred shipyard rats are likely to be greater than the genetic differences among an inbred population of laboratory rats. All things being equal, the cumulative quantal dose–effect curve for the shipyard rats is therefore likely to be shallower than that for the inbred rats, because a wider range of doses will probably be required to produce the specified endpoint, death, due to the greater genetic variety within the shipyard population.

8. The mean dose is defined as the average dose administered to a test population. Like the median dose, the mean dose is a statistic, not an actual dose. When the frequency histogram is a perfect bell-shaped curve, i.e., a perfect normal (Gaussian) distribution, the mean dose is equal to the median dose. A major reason for deviations from this ideal is the relatively small size of real-world test populations. A second reason is that the administered doses may be inadvertently skewed toward the high or low end of the frequency distribution. A third possibility is that the frequency distribution for a particular toxic effect may not be completely normal.

9. Determine the dose of test chemical that causes a 50% decrease in red blood cells:
    a. Figure 3.2a: Draw a horizontal line from 50% on the y-axis until it intersects the graded dose–effect curve, then drop a perpendicular line from the point of intersection to the x-axis. Read the dose that causes a 50% decrease in red blood cells directly at the point of intersection on the x-axis. The result is about 9 mg/kg based on a visual estimate. A more accurate result may be obtained by measuring the distance on

the x-axis from 0 to the point of intersection, dividing it by the distance from 0 to 10, and multiplying the resulting fraction by 10. The result obtained by using the graph in Figure 3.2a can be checked by inspection of Table 3.2. According to Table 3.2, a dose of 10 mg/kg causes a 51.9% decrease in red blood cells. Thus, 9 mg/kg is a plausible result.

b. Figure 3.2b: The same procedure is followed as described above for Figure 3.2a for finding the dose corresponding to the ED_{50} on the x-axis. Because Figure 3.2b is a semilogarithmic plot (the x-axis is scaled in logarithmic units, while the y-axis is scaled in linear units), it is harder than in Figure 3.2a to estimate the dose by interpolating between 5 and 10. For greatest accuracy, use a ruler to measure two distances on the logarithmic scale of the x-axis: The distance between 1 and the point of intersection (call it a) and the distance between 1 and 10 (call it b). Because the x-axis is scaled in logarithmic units, the dose in question is equal to $10^{a/b}$.

c. Equation (3.1):

Effect magnitude = Maximal effect [Dose/(Dose + EC_{50})]

Rearranging this equation,

Dose = EC_{50}/[(Maximal effect/Effect magnitude) – 1]

The maximal effect is a 70% decrease; the effect magnitude is a 50% decrease; and the ED_{50} is 3.5 mg/kg. Therefore,

Dose = (3.5 mg/kg)/[(70%/50%) – 1] = 8.75 mg/kg

10. Compute incidence of anemia at a test chemical dose of 30 mg/kg body weight:
    a. Equation (3.2):

Effect frequency = Maximal frequency [Dose/(Dose + Median dose)]

Effect frequency = 100% [30 mg/kg/(30 mg/kg + 63 mg/kg)]

Effect frequency = 100% (30/93) = 32%

b. Figure 3.1b: Because the x-axis of Figure 3.2b is scaled in logarithmic units, it is difficult to estimate visually where a dose of 70 mg/kg lies on the x-axis. For greater accuracy, its location on the x-axis may be calculated. It is important to recognize that, on the logarithmically scaled x-axis, the distances between doses correspond to the logarithms of the doses. The numbers on the x-axis, i.e., the doses, are the antilogs of the distances on the x-axis. For example, the distance between 1 mg/kg and 2 mg/kg is (or should be, if the figure is drawn accurately) equal to 30% or 3/10 of the distance between 1 mg/kg and 10 mg/kg. Why? Because $10^{0.3} = 2$. Thus the dose, 2 mg/kg, is the antilog of the fractional distance on the x-axis, i.e., 30% of the distance corresponding to one logarithmic cycle (a logarithmic cycle is any multiple of 10, i.e., 1 to 10, 10 to 100, 100 to 1000, etc.). A dose may be located on the x-axis in three steps: First, divide the dose by 10, 100, 1000 or any other multiple of 10 as required to reduce it to a number between 1 and 10. (If the dose is less than 1, e.g., 0.7, then multiply it by 10, 100, 1000 or any multiple of 10 as required to increase it to a number between 1 and 10.) Second, determine the logarithm of the dose after reducing (or increasing) it to a number between 1 and 10. Third, multiply the logarithm times the actual physical distance of the logarithmic cycle on the graph. In Figure 3.1b, in order to locate the dose of 30 mg/kg on the x-axis, first divide 30/10 = 3. Second, find the logarithm of 3, which is 0.48. Third, measure the physical distance of one logarithmic cycle on the graph. In the manuscript of this book, the physical distance on the x-axis between 10 mg/kg and 100 mg/kg (as well as between 1 mg/kg and 10 mg/kg, 100 mg/kg and 1000 mg/kg, etc.) is 29 mm. Therefore, the dose in question, 30 mg/kg, is located 0.48 × 29 mm = 14 mm to the right of 10 mg/kg. A line drawn perpendicular to the x-axis from this point intersects the curve in Figure 3.2b at a point corresponding to a value on the y-axis of 30%. This result agrees well with the result of 32% calculated under 10a, but not perfectly—which is not surprising, given potential sources of error in both methods.

SUGGESTED READING

Hughes, W. W. 1996. *Essentials of environmental toxicology*. Washington, DC: Taylor & Francis.
Klaassen, C. D. 2001. *Casarett & Doull's toxicology: The basic science of poisons*, 29–41. New York: McGraw-Hill.

4 Human Populations at Risk

4.1 INTRODUCTION

The first three chapters of this book introduced the conceptual basis for assessing toxic chemical risk. Risk depends on a combination of chemical toxicity and exposure (Chapter 1). Exposure can occur when toxic chemicals released into the environment undergo partitioning and advective transport that bring them into contact with biological receptors (Chapter 2). The dose–effect relationship (Chapter 3) provides a basis for understanding and managing toxic chemical risk. We turn now from the conceptual to the practical: How can the dose–effect relationship be used to assess and manage toxic chemical risk? In this chapter we look at epidemiological studies as one approach to identifying chemicals that cause disease in human populations.

4.2 LAW AND LOOPHOLES

The goal of regulatory toxicology is to protect citizens from the risks posed by toxic chemicals by mediating between businesses that profit from the use of toxic chemicals in their products, on the one hand, and members of the public who use those products, on the other. There are roughly 80,000 different chemical compounds currently reported as being used in commercial products in the United States. The United States is not unique: A similar variety of chemicals is present in goods marketed in other countries, as well. Each year, several hundred new chemical compounds are created by industry and brought to market. What safeguards are in place to ensure that citizens are not harmed by the chemicals to which we are all exposed?

A number of laws have been passed by Congress that are intended to provide safeguards against toxic chemical risk (Table 4.1). While these laws have provided real benefits, protecting public health and the environment from toxic chemical risk remains a work in progress. The Toxic Substances Control Act (TSCA) is a case in point. Passed by Congress in 1976, TSCA requires that newly created chemicals undergo toxicity testing before they are brought to market. However, TSCA exempted from toxicity testing all chemicals marketed prior to December 1979. The exempted chemicals were placed on a list called the TSCA inventory, approximately 62,000 chemicals in all. There are no toxicity data on those 62,000 chemicals (except for data that might have been collected under other federal statutes)—a toxicological black hole that makes it essentially impossible to evaluate their health and environmental risks. In addition to a lack of toxicity data on chemicals marketed prior to 1979, some 85% of new chemicals reaching the market are not actually tested

53

TABLE 4.1
Major Laws in the United States that Protect Human Health and the Environment against Hazardous Chemicals

Federal Agency	Regulated Entity	Federal Laws
EPA	Air pollutants	Clean Air Act 1970, 1977, 1990
	Water pollutants	Federal Water Pollution Control Act 1972, 1977
	Drinking water	Safe Drinking Water Act 1974, 1996
	Pesticides	Fungicides, Insecticides, and Rodenticides Act (FIFRA) 1972
		Food Quality Protection Act (FQPA) 1996
	Ocean dumping	Marine Protection Research, Sanctuaries Act 1995
		Ocean Radioactive Dumping Ban Act 1995
	Toxic chemicals	Toxic Substances Control Act (TSCA) 1976
	Hazardous chemicals	Resource Conservation and Recovery Act (RCRA) 1976
	Abandoned hazardous wastes	Superfund (CERCLA) 1980, 1986
CEQ	Environmental impacts	National Environmental Policy Act (NEPA) 1969
OSHA	Workplace	Occupational Safety and Health (OSH) Act 1970
FDA	Foods, drugs, and cosmetics	FDC Acts 1906, 1938, 1962, 1977
		FDA Modernization Act 1997
CPSC	Dangerous consumer products	Consumer Product Safety Act 1972
DOT	Transport of hazardous materials	THM Act 1975, 1976, 1978, 1979, 1984, 1990

Source: Reprinted with permission from Curtis Klaasen, *Casarett & Doull's Toxicology: The Basic Science of Poisons*, 6th ed. (New York: McGraw-Hill, 2001), 86.

Note: Abbreviations of enforcing agencies: EPA, Environmental Protection Agency; CEQ, Council for Environmental Quality (now Office of Environmental Policy); OSHA, Occupational Safety and Health Administration; FDA, Food and Drug Administration; CPSC, Consumer Product Safety Commission; DOT, Department of Transportation.

---

for health effects, TSCA notwithstanding. The TSCA inventory currently contains about 83,000 chemicals. Of these, it is estimated that only on the order of 7% have been fully characterized with respect to their toxicity. The Environmental Protection Agency (EPA) has required testing on 200 chemicals since TSCA was passed in 1976, and it has banned just five chemicals under TSCA. According to some sources, the consensus in and out of government is that while TSCA represented a major conceptual advance in protecting the public from toxic chemical risk, in practice it has been largely ineffectual. Indeed, it is sometimes referred to mockingly by its critics as the "Toxic Substances Conservation Act" (Schapiro 2007).

The European Union (EU) has recently emerged as a global leader in identifying and reducing risks from toxic chemicals. In response to growing public concern based on studies showing convincingly that EU citizens, like Americans, have elevated body burdens of dozens of chemicals that might pose long-term risks of cancer and reproductive disorders (U.S. Department of Health and Human Services

2005; World Wildlife Fund 2005), the EU issued a directive called RoHS (Removal of Hazardous Substances) that banned six chemicals from electrically powered devices effective July 1, 2006: mercury, cadmium, lead, hexavalent chromium, and two members of a class of flame retardants called polybrominated biphenyls. In addition, the EU has initiated a program called REACH (Registration, Evaluation, Authorization and Restriction of Chemicals) with the goal of providing toxicity data on each and every chemical in commerce in the European Union. The EU's scientific committees, which are made up of experts from all 25 EU countries, began reviewing chemicals under REACH in June 2007. The scientific committees are compiling a list of substances of very high concern, and manufacturers will soon be required to limit or remove these substances from their products unless they demonstrate that the risks they pose can be adequately controlled. Like the European Union, Canada has taken steps to screen all chemicals in commerce and to restrict the use of those determined to be toxic (Chapter 10).

The new initiatives in Canada and the European Union and TSCA's groundbreaking approach in the 1970s attempt to address the same fundamental problem: Toxic chemicals have become a central feature of the national and global economy, and it is imperative that they be regulated to protect human health and the environment. The question is, how should regulation be structured? From the standpoint of business, the question of regulation might be formulated as follows: How can adequate protection be maintained without damaging profitability? From the standpoint of the general public, the emphasis might be reversed: How can adequate economic progress be maintained without endangering public health?

For those with gallows humor, the importance of having an effective regulatory structure is underlined by two so-called laws of human experience. Murphy's Law states that if something can go wrong, it generally does. Examples of Murphy's Law from the realm of regulatory toxicology are that toxic chemicals may not always be used as intended nor disposed of as mandated—with the result that there are now literally tens of thousands of uncontrolled hazardous chemical waste sites scattered around the country. The other law is the "law of unintended consequences," which states that things can and do happen that were not anticipated at all. Examples abound in the world of toxic chemical risk: Lead that was once used to enhance the qualities of paint now impairs the cognitive development of thousands of children who live in houses built before lead paint was banned and accidentally eat paint chips. DDT has been banned in the United States because it turned out to be a potent reproductive toxicant in birds, particularly one of its transformation products, DDE, which causes eggshell thinning. Freon and other chlorofluorocarbons that were marketed as refrigerants have been found to degrade the earth's stratospheric ozone shield, increasing the incidence of skin cancer. Not only marketed chemicals, but chemicals that are generated as waste products may also enter the environment and cause unintended consequences to human and ecosystem health. Perhaps the premier example is the generation of carbon dioxide from the combustion of fossil fuels, now understood to be the leading cause of global warming. Crafting a truly effective regulatory system for managing toxic chemical risk without unduly slowing economic development appears to be one of the great challenges in creating a sustainable global society.

4.3 AFTER THE FACT

The cornerstone of any regulatory system consists largely of toxicity data from whole animal studies that make it possible to predict, with a reasonable degree of confidence, the risks a chemical poses to human health and the environment (Chapter 5). Predictions from animal models are used to manage chemical risk prospectively. For example, government may impose restrictions on a chemical's distribution, require that the manufacturer provide labels describing adverse effects, provide guidelines for safe disposal, and ultimately ban the chemical if the risks it presents are judged to outweigh its usefulness. While toxicity studies on animals are extremely informative and absolutely essential for protecting human health, they are not infallible. It is still possible for chemicals to get into the environment that are not supposed to be there, and for chemicals and/or their transformation products to turn out to be more harmful than predicted. How does society know when a chemical that was predicted to be safe is actually causing harm—when a chemical genie has gotten out of its bottle?

In contrast to the regulatory approach of preventing harmful chemical exposure before it occurs, which is a proactive approach, epidemiology takes a retrospective approach to risk. It asks whether chemical exposure and harm have already occurred. This may seem like closing the barn door after the horse has gotten loose. However, the role of human epidemiological studies in the management of risk is to complement, not replace, predictions based on extrapolations from toxicity testing in animals. For chemicals that have been insufficiently characterized, epidemiological studies are the only means of uncovering toxic effects. They are capable of showing whether humans in fact are—or are not—being harmed accidentally by toxic chemicals.

The specific question that epidemiologists ask is whether a population thought to be exposed to a toxic chemical is manifesting a higher than normal incidence of a specific toxic effect predicted for that chemical. Stated another way, epidemiologists have to decide whether a population of humans that is believed to be at risk does, in fact, have an elevated frequency of a characteristic toxic effect. How do epidemiologists make this decision? The short answer is that they compare the frequency, or incidence, of the characteristic toxic effect in the at-risk population to the frequency of the characteristic toxic effect in a control population that is believed not to be at risk, i.e., not to be exposed. A significant difference in disease frequency constitutes evidence that the at-risk population is, in fact, suffering from an elevated incidence of chemically induced disease.

Epidemiological studies, which are basically statistical comparisons between two different human populations, are strongest when the populations are large and the incidence of disease in the at-risk population, i.e., the frequency of the specified toxic endpoint, is high. They are weakest when study populations are small and the toxic endpoint frequency is low. To gain a better appreciation of the strengths and weaknesses of the epidemiological approach to identifying accidental chemical disease, let us take a closer look at the statistical reasoning on which it is based.

4.4 THE NULL HYPOTHESIS AND STATISTICAL POWER

When epidemiologists compare two human populations, one defined as being at risk and the other defined as the control, they begin by hypothesizing that there is no difference in disease frequency between the two populations. They then collect data to decide whether their hypothesis is correct or incorrect. The hypothesis of no difference between two populations is called the null hypothesis. The null hypothesis is accepted if it is decided that there is no difference between the two populations, and it is rejected if it is decided that there is a difference. There is a finite probability of committing an error and rejecting the null hypothesis when it should be accepted and of accepting the null hypothesis when it should be rejected. The decision to accept or reject the null hypothesis is associated with a specified level of statistical confidence in the data. For example, if the null hypothesis is rejected at the 0.95 confidence level, there is a 95% chance that the decision is correct (i.e., that there really is a difference between the study and control populations) and a 5% chance that the decision to reject the null hypothesis is erroneous (i.e., that there really is no difference between the two populations).

The statistical significance for a calculated confidence level depends on a study's statistical power. The greater the statistical power, the greater the probability that the difference—or lack of difference—between the study and control populations is statistically significant. Clearly, having more statistical power is a very desirable state of affairs for epidemiologists and toxicologists. There are two ways to increase statistical power. One is to increase the number of samples drawn from the populations being compared. The larger the number of samples, the smaller the difference in disease frequency between the two populations that can be detected. The second way that statistical power increases is when the incidence of disease in the at-risk population is high compared to the control population. Statistical power is greatest, then, when sample sizes are large and the incidence of disease in the at-risk population is high.

Because epidemiology and toxicology use statistics to estimate and compare disease frequencies, the issue of statistical power arises again and again. In human epidemiology, statistical power is usually increased by increasing the sample sizes drawn from the at-risk and control populations being studied. Increasing sample sizes reduces the probability that a chemically induced disease will go undetected. The largest human epidemiological studies are capable of detecting differences in disease frequency, also referred to as excess risk, in the 0.1% range, or 1 in 1,000. However, such large studies are expensive and resource-intensive, and therefore they remain the exception. Most epidemiological studies are capable of detecting approximately 1% excess risk, or 1 in 100 (Figure 4.1).

In animal toxicity testing (Chapter 5), large populations of test animals are avoided because they are prohibitively expensive to maintain. Instead, the second strategy for increasing statistical power, increasing the frequency of chemically induced disease, is pursued by administering high doses of chemicals to the test population of animals. High doses increase the frequency of chemical disease and hence the statistical power of the toxicity test. As a result, statistically significant results can be obtained using smaller test populations. Unfortunately, high doses often introduce scientific

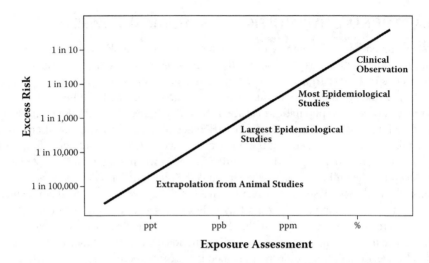

FIGURE 4.1 Ability of epidemiological studies to detect toxic chemical risk. The graph shows the excess risk that can be detected as a generalized function of exposure. The higher the exposure, the greater the risk. Most epidemiological studies are able to detect excess risk when exposure is high enough to produce an approximately 1% increase in risk. (Reprinted with permission from Ian Pepper, Charles Gerba, and Mark Brusseau, eds., *Pollution Science* [San Diego: Academic Press, 1996] 348.)

ambiguity in the interpretation of test results. In particular, the biological significance of high-dose exposure and its relevance to the much lower exposures typically encountered by humans can vary considerably with different chemicals.

4.5 PROOF OF CAUSATION

In human epidemiology, a decision to reject the null hypothesis means that the study population defined as being at risk has a significantly higher incidence of a specified toxic effect than the control population. Such a finding constitutes strong evidence of a link between chemical exposure and disease. Factors that might contribute to differences in disease frequencies but that are not related to the chemical of concern, collectively known as confounding factors, are excluded by well-designed epidemiological studies. Examples of confounding factors are family (genetic) history, socioeconomic differences, medical conditions such as poor nutrition, compromised immune system function, preexisting disease conditions, and lifestyle choices, e.g., cigarette smoking and alcohol consumption.

Despite the power of well-designed epidemiological studies to detect differences in disease frequencies in human populations, it is important to keep in mind that statistical significance is not the same as toxicological significance. To make a convincing connection between chemical exposure and disease, one or more additional kinds of evidence are needed. First, there must be evidence that the at-risk population was actually exposed to the chemical of concern. Second, the chemical should obey the dose–effect relationship, i.e., the incidence of the toxic effect of concern should, if possible, be shown to increase as a function of exposure, either in the at-risk population, in a population of test animals, or both. Third, it is important to know from animal toxicity testing or from previous accidental exposures of humans whether the chemical is capable of producing the observed toxic effect at the exposure levels that have occurred. One or more of these three additional pieces of evidence—exposure, dose–effect, and toxicological plausibility—when coupled with the results of an epidemiological study, are usually enough to persuade government to take regulatory action in the interest of protecting public health. For example, in the Yusho ("[rice] oil disease") poisoning incident in 1968, polychlorinated biphenyls (PCBs) and polychlorinated dibenzofurans (PCDFs) leaked from machinery at a rice oil factory in Japan. The contaminated oil killed 400,000 chickens and sickened 14,000 people before it was discovered. Symptoms included skin sores, irregular menstrual cycles, and impaired immune responses as well as fatigue, headache, and cough. Japanese scientists combined epidemiological studies with investigations of exposure and dose–effect relationships in a model of toxicological detective work that eventually led to a ban on industrial uses of PCBs in Japan and other countries, including the United States.

In addition to exposure, dose–effect, and biological plausibility, a fourth type of evidence can occasionally be advanced to help clinch the case against a suspect chemical: its molecular mechanism of action, the toxicological equivalent of the smoking gun. Knowing how a chemical performs its harmful action constitutes, in principle at least, proof positive of its guilt. There are numerous examples of chemicals, especially toxicants with a long history of human exposure and concern such as lead, arsenic, mercury, cyanide, carbon monoxide, and various snake, plant, and fungal toxins for which molecular mechanisms of action are well understood. The toxicities of most therapeutic drugs are also known in considerable, though not always complete, molecular detail. Yet for most diseases and most chemicals, the molecular connection can be difficult to make; as a result, a chemical's mechanism of action rarely serves as a basis for regulating its use. Three reasons stand out.

First, there is usually not a large enough body of research from which to draw conclusions about molecular causality. Second, the etiology (origin) of a disease may depend on multiple steps, one or more of which involve toxic chemicals and others of which do not involve toxic chemicals at all. Cancer is a case in point (Chapter 7). Disease processes like cancer, which involve multiple steps and occur over a long period of time, make it hard to assess the precise role of a toxicant that may be involved in only one step of many. A third obstacle to proving the impact of a suspect chemical on public health is that exposure may be to a mixture of chemicals rather than to a single chemical. Cigarette smoke, a complex mixture of scores of chemicals, illustrates this point. Polychlorinated biphenyls, a family of 209 congeneric (related but structurally distinct) molecules, are another example of a complex chemical mix. When chemical disease results from exposure to a heterogeneous mixture of chemicals and not to a single chemical compound, establishing molecular causality becomes an extremely difficult proposition. Examples of the kinds of obstacles involved are, first, the need to purify the chemicals in the mixture and investigate them, one by one; and second, the need to consider the possibility that two or more chemicals act in concert to trigger the disease rather than one chemical in the mixture acting alone.

For these and other reasons, identifying the molecular smoking gun on which to convict a chemical—or a chemical mixture suspected of harming an at-risk population—can be an elusive goal. Fortunately, taking reasonable action to protect public health almost never depends on an understanding of molecular causality. For example, the U.S. government banned PCBs on the basis of epidemiological and toxicological evidence, even though the molecular mechanisms of PCB toxicity were, and still are, not known in detail. If regulation of toxic chemicals depended on detailed knowledge of molecular cause and effect, most chemicals would not be subject to regulation.

4.6 DESIGNING AN EPIDEMIOLOGICAL STUDY: COHORT VS. CASE CONTROL

There are two basic approaches to designing an epidemiological study to investigate accidentally induced chemical disease: cohort and case control. In a cohort study, the at-risk population is defined as one that is known to have been exposed to a chemical of concern, and the frequency of disease is compared with that of a control population that has not been exposed. More than one disease may be tracked in a cohort study, depending on the known or suspected toxicities of the chemical of concern. Elevation of disease frequency, i.e., rejection of the null hypothesis at an acceptable level of statistical confidence, constitutes presumptive evidence of a causal link between exposure and disease. The causal connection can be strengthened by evidence of a dose–effect relationship, as discussed in Section 4.5.

There are two types of cohort study design, retrospective and prospective. A retrospective cohort study looks at the past disease frequency in a population that was exposed prior to the onset of the study. An example of a retrospective cohort study is one that compares the incidence of lung cancer from 1985 to 2000 in a group of people who smoked two packs of cigarettes a day (the cohort) with the lung cancer incidence in a control group of nonsmokers over the same period. Exposure and disease are both in the past tense in a retrospective cohort study. In a prospective cohort study, exposure to a toxic chemical has also occurred prior to the initiation of the study, but the incidence of disease is tracked, not in the past, but in the present and going forward into the future as symptoms continue to emerge in the exposed population. Perhaps the largest prospective cohort study in the world is the continuing investigation of the survivors of the atomic bomb attack on Hiroshima, Japan, in 1945. Both the survivors and their descendants have been included in the study because of the possibility that atomic radiation caused heritable damage to the genes of people who survived exposure at ground zero.

In a case-control study design, the point of departure is an at-risk population that is defined, not as one that is known to have been exposed, but as one that is suspected of having an elevated disease frequency (the case). The disease frequency is compared with that of a sample drawn from a second population that is thought not to be at risk (the control). Elevation of the disease frequency in the study group drawn from the "case" population compared with the study group drawn from the "control" population, i.e., rejection of the null hypothesis at an acceptable level of statistical confidence, constitutes presumptive evidence of an increased risk of disease.

If, in a case-control study, there is a statistically significant difference between the at-risk and control populations but no evidence of exposure, the result may not be convincing, particularly if the difference is small. In order to be convincing, a case-control study needs to establish that the at-risk population has in fact been exposed to a particular toxic chemical, what the magnitude and duration of the exposure was, and whether the disease is consistent with the chemical's known toxicity.

Two popular movies, *A Civil Action* and *Erin Brockovich*, both based on real-life stories, illustrate the strengths and limitations of the case-control study design. In *A Civil Action*, an elevated incidence of health problems in East Woburn, Massachusetts, including leukemia, which has taken the lives of several children, is documented in an initial small case-control study. The families decide to sue, bringing what is known as a "toxic tort" lawsuit against the two industries that they believe are responsible for poisoning their water supply. The plot then turns on the efforts of the families' lawyer, Jan Schlichtman, to obtain conclusive evidence of exposure to a chemical that could have caused the leukemia and other health problems—evidence that can stand up in court.

In a toxic tort, the responsible party must be shown with a reasonable degree of certainty to have caused the alleged harm. The required evidence is of two kinds. First, the alleged perpetrators, in this case the industries, must be shown to have dumped a chemical in the environment that can cause leukemia. Second, a plausible pathway of exposure must be traced from the chemical dumpsite to East Woburn residents' water supply on the basis of generally accepted principles of chemical fate and transport in the environment (Chapter 2). Schlichtman wins a partial victory against one of the industries, but the expense of retaining expert witnesses bankrupts his law firm. When the court case is over, he shares the evidence he has gathered with the U.S. Environmental Protection Agency. The EPA eventually uses its regulatory authority to require the industries to clean up some of the contamination.

A Civil Action illustrates an important point about regulating toxic chemical risk: The EPA's authority to require a cleanup is not based on a successful toxic tort lawsuit demonstrating harm by the alleged perpetrators to an individual person or persons. Rather, it is based on levels of toxic chemicals in soil and groundwater that are predicted from animal toxicity tests to be harmful to human health. This fundamentally new principle, that polluters be held liable on the basis of concentrations of toxic chemicals in the environment and not on the basis of toxic tort lawsuits in a court of law, was embodied in the Comprehensive Environmental Response, Compensation and Liability Act (CERCLA). Also known as the Superfund Law, this act was passed by Congress in 1980. Under CERCLA, federal and state regulatory agencies do not have to prove a cause-and-effect relationship between a specific instance of environmental contamination, on the one hand, and chemical diseases such as leukemia in East Woburn, on the other. Rather, the fact that environmental samples are found to contain toxic chemical concentrations that are known to be harmful on the basis of scientifically credible laboratory studies is sufficient to trigger the government's statutory authority to require a cleanup. Regulation based on a body of scientific evidence instead of the successful prosecution of toxic tort cases was a groundbreaking concept in 1980, and it was emulated around the world. The new regulatory initiatives in Canada and the European Union effectively continue CERCLA's trajectory of science-based protection of public health.

A Civil Action brings out the difference between protecting public health by regulating levels of toxic chemicals in the environment, on the one hand, and, on the other, protecting public health by suing businesses in order to make it costly for them to contaminate the environment. The book also illustrates the difference between scientific evidence, which encompasses probability and hypothesis and which can serve as a statutory basis for regulatory action even if harm to individual humans cannot be proven, and evidence in a court of law, which needs to be concrete and linked to the actions of specific individuals in order to reach a verdict. Another take-home lesson from *A Civil Action* is that it is not a good idea for the public to rely on lawsuits for protection from toxic chemical risk, because the legal profession's bar for evidence to convict a defendant in a court of law is very high. Science-based environmental statutes and regulations do a better job of protecting public health—always provided, of course, that the laws and regulations are reasonable, workable, and enforced.

Yet even when they are effective, environmental laws and government regulatory agencies are neither omniscient nor infallible. Sometimes, they end up missing something big, people get hurt, and a lawsuit turns out to be not only appropriate, but actually winnable. In the movie *Erin Brockovich*, the title character is a divorced single mother with grit, good looks, and a talent for toxicological epidemiology. She gathers evidence showing that residents near a Pacific Gas and Electric Company (PG&E) facility in California are suffering from a spectacularly elevated incidence of cancer and other diseases. She also discovers that the PG&E facility is dumping hexavalent chromium, a chemical known from animal toxicity studies to be capable of causing the kinds of diseases seen in the people who live near the plant. (It is also one of the six chemicals banned by the European Union's RoHS directive.) In addition to convincing epidemiological evidence, i.e., the high incidence of disease in the exposed population, Brockovich gets some lucky breaks that eluded Jan Schlichtman in *A Civil Action*. Most importantly, she obtains an internal memo from a former PG&E employee constituting legal proof that the company knew it was dumping a dangerous chemical in the environment. In terms of a toxic tort lawsuit, it all adds up to a perfect legal storm for PG&E. Given the scientific and legal evidence arrayed against it, PG&E agrees to a multimillion-dollar settlement rather than going to court. *Erin Brockovich* is a nice story of the triumph of justice over evil. However, the science and law on which it is based—an ironclad case-control study and incontrovertible legal evidence of wrongdoing—are the exception, not the rule. Usually, the scientific and legal evidence in toxic tort cases is more tenuous, as in *A Civil Action*.

The two classes of epidemiological studies, case-control studies and cohort studies, are both based on the dose–effect relationship, i.e., the relationship between the magnitude of chemical exposure and the frequency of disease. The two study designs approach the dose–effect relationship from different directions, case-control studies from the effect side and cohort studies from the exposure side. As might be expected, they meet somewhere in the middle. In fact, case-control studies may evolve into cohort studies once the source of exposure is known. This happened in *Erin Brockovich*, which began with a case-control study of a suspected cancer cluster, moved to the identification of hexavalent chromium as the probable causative agent, and then, with hexavalent chromium exposure as a new point of departure, ended as a cohort study of residents near the PG&E facility who were known or suspected to have been exposed.

4.7 LEVEL I, II, AND III EPIDEMIOLOGICAL STUDIES

Epidemiological studies are divided into three general levels, depending on the kinds of data that are used in the study. Level I studies, which are relatively inexpensive and are intended to generate hypotheses for further testing, draw on data from vital records in the public domain. Examples are the cancer registry, birth registries, and death certificates. Depending on the outcome of a level I study, a level II study may be indicated. For example, an increase in the incidence of cancer in a city or a region might trigger a level II study.

A level II study gathers data that are not available in public records. The kinds of data depend on whether the study design is case control or cohort. A level II case-control study typically investigates pathways of exposure for the at-risk population. An investigation of exposure includes interviews with individuals in the at-risk population about their length of residence in the presumed impact area, dietary habits, workplace conditions, and places where children play. It may also include qualitative or semiquantitative modeling of the environmental fate and transport of chemicals of concern (Chapter 2). Samples of body fluids and tissues such as blood, urine, hair, or nails may be analyzed for evidence of exposure to bioaccumulative chemicals such as arsenic, mercury, or PCBs. An effort is made to identify and exclude possible confounding factors. *Erin Brockovich* and *A Civil Action* are examples of level II case-control studies.

A level II cohort investigation is based on known exposure and focuses on symptoms of disease. Individuals in the exposed population are interviewed in detail about their medical histories to determine whether their symptoms coincide with toxic effects associated with the chemical of concern. Potential confounding factors are also investigated. Clinical drug trials and investigations of the health effects of workplace exposures to toxic chemicals are examples of level II cohort studies.

Level III applies only to prospective cohort studies, which involve decades-long medical follow-ups of large study populations. For example, the Nurses' Health Studies have been investigating possible health effects of oral contraceptives by following cohorts totaling 238,000 female nurses since 1976. The two cohorts of atomic bomb survivors and their descendants, one at Hiroshima and the other at Nagasaki, Japan, have been followed continuously since shortly after the end of World War II.

Epidemiological studies of disease frequency are capable of providing solid evidence of chemically induced disease in human populations. However, evidence of a toxicological link to chemical exposure is also needed to corroborate the epidemiological findings and give them a biological context. Toxicological evidence comes overwhelmingly not from data on accidental exposures of humans, but from controlled testing in laboratory animals, the subject of the next chapter.

STUDY QUESTIONS

1. Describe the basic design of an epidemiological study.
2. What is the null hypothesis? How is it used to investigate toxic chemical risk in human populations?
3. Describe the features that confer statistical power on a human epidemiological study. Discuss the ways that statistical power can be increased.
4. Describe the kinds of toxicological evidence that serve to substantiate and strengthen epidemiological (statistical) evidence of chemical disease.
5. Define what is meant by a cohort study. Define what is meant by a case-control study. What are the advantages and disadvantages of each type of study with respect to protecting the public from toxic chemical risk?
6. How does a prospective cohort study differ from a retrospective cohort study?
7. Describe a level I and a level II case-control study. What are the major differences between them?
8. Was *A Civil Action* based on a case-control study or a cohort study? Explain.
9. The law firm in *Erin Brockovich* succeeded in their toxic tort lawsuit against a polluting industry, while the law firm of Jan Schlichtman failed in *A Civil Action*. Discuss the scientific and legal reasons for these different outcomes, including the nature of scientific evidence and the requirements for legal proof in a court of law.
10. Discuss the pros and cons of protecting public health from toxic chemical risk using environmental laws such as CERCLA, on the one hand, and using toxic tort lawsuits as exemplified by *A Civil Action* and *Erin Brockovich*, on the other.

ANSWERS TO STUDY QUESTIONS

1. An epidemiological study compares the incidence of disease in an at-risk population to the incidence of disease in a control population. Due to limitations of time and resources, the actual study populations are subsets of the total at-risk and control populations.
2. The null hypothesis states that there is no difference between the at-risk population and the control population in the rate or incidence of disease. Investigators perform statistical analyses of study results to decide whether to reject the null hypothesis (i.e., there is a significant difference in disease incidence) or accept the null hypothesis (i.e., there is not a significant difference in disease incidence).
3. Human epidemiological studies gain statistical power when sites of the samples drawn from and control populations increase in the study and when the incidence of disease in the study population increases. The only way to increase statistical power in human epidemiological studies is to increase sample sizes. In toxicity testing in animals, which are essentially small-scale epidemiological studies under controlled laboratory conditions, statistical power can be increased by increasing the doses of toxic chemicals and, hence, increasing the incidence of disease in the test population. Unfortunately, while

this strategy increases statistical power, it also introduces biological ambiguity, because humans are almost always exposed to low doses, not high doses, and it is often an open question as to whether the body's response to low doses is the same or different than its response to high doses of toxic chemicals.
4. Rejection of the null hypothesis provides prima facie statistical evidence of increased disease incidence in an at-risk population; however, it does not constitute conclusive proof. One or more of several other lines of evidence are needed to buttress the statistical finding: (a) evidence that the at-risk population has been exposed to the toxic chemical; (b) evidence from toxicity testing in animals or from accidental human exposures that the chemical produces the disease symptoms observed in the at-risk population; (c) evidence that the molecular mechanism of action is consistent with the observed disease symptoms; and/or (d) evidence that the incidence of chemical disease conforms to a dose–effect relationship.
5. A cohort study is an epidemiological study in which an at-risk population, a "cohort" is known to have been exposed to a chemical of concern, and symptoms of disease are investigated in the exposed population and compared to the incidence of disease symptoms in an unexposed control population. A case-control study is one in which the at-risk population, the "case," is suspected of having an elevated incidence of disease and is investigated by comparing its rate of disease with that of a control population. A public health benefit of a case-control study is that it can uncover toxic effects of a chemical that were previously unknown or underappreciated. A public health benefit of a cohort study is that it can provide definitive documentation of toxic effects on an exposed population.
6. A cohort study may be retrospective, i.e., exposure and disease occurred prior to the onset of the study. Alternatively, a cohort study may be prospective. In this case, exposure occurred before the study began, but disease incidence is investigated, not in the past, but going forward into the future.
7. A level I case-control study utilizes vital records in the public domain such as the cancer registry, birth registry, or death certificates to compare the rate of disease in a control population with the rate of disease in a putative at-risk population. Level I studies are relatively inexpensive and are generally designed to generate hypotheses for further testing in level II case-control studies. A level II case-control study gathers data that are not available in the public domain and may include interviews with individuals in the at-risk population, samples of body fluids and tissues, and modeling of the environmental fate and transport of the chemical(s) of concern.
8. Based on a true story, *A Civil Action* turns on a case-control study: The incidence of leukemia appears to be elevated in a particular area in East Woburn, Massachusetts. The book chronicles the efforts of the residents' lawyer, Jan Schlichtman, to prove that the apparent leukemia cluster is a bona fide "case" of poisoning of an at-risk population, and that it was caused by the illegal disposal practices of two nearby chemical industries. Schlichtman produces a great deal of evidence, and while it is enough to persuade many scientists of a connection between Woburn residents' leukemia and the industries' disposal

practices, it is not enough to satisfy the legal criterion of guilt beyond a reasonable doubt. He loses his case and bankrupts his law firm.

9. Also based on a true story, *Erin Brockovich*, like *A Civil Action*, turns on a case-control study of an elevated rate of disease among people who live near an industry suspected of dumping toxic chemicals in the environment. Where Jan Schlichtman failed to make a convincing legal case, Erin Brockovich succeeds, for three reasons: First, the incidence of disease in the "case" population is very high and therefore convincing from a statistical standpoint; second, she obtains a confidential internal memo discussing the company's illegal dumping of hexavalent chromium; and third, the victims' symptoms are consistent with the known toxic effects of hexavalent chromium. Thus, the scientific evidence is strong: elevated incidence of disease symptoms consistent with a specific chemical of concern known to have entered the nearby environment. The legal evidence is also strong: an incriminating memo showing intent to dispose of the chemical of concern illegally. The company settled out of court.

10. Toxic tort lawsuits are difficult to prosecute because the scientific and legal bars for proof are high. They are useful as a deterrent to polluters and as a way to raise public awareness of problems and abuses. However, on the broad scale needed to safeguard public health *vis-à-vis* the pervasive presence of chemicals in our lives, toxic torts must be regarded on balance, as relatively ineffectual. Federal and state statutes that use the best available scientific evidence to set and enforce levels of toxic chemicals in the environment are able to provide broader and better protection of public health than toxic torts.

REFERENCES

European Union. RoHS compliance. http://www.rohs.eu/english/index.html
Harr, J. 1995. *A civil action*. New York: Vintage Books.
Harvard University. Nurses' health study. http://www.channing.harvard.edu/nhs/
Klaasen, C. D. 2001. *Casarett & Doull's toxicology: The basic science of poisons*. 6th ed. New York: McGraw-Hill.
Pepper, I. L., C. P. Gerba, and M. L. Brusseau, eds. 1996. *Pollution science*. San Diego: Academic Press.
Schapiro, M. 2007. *Exposed: The toxic chemistry of everyday products and what's at stake for American power*, pp 124–158. White River Junction, VT: Chelsea Green Publishing.
U.S. Department of Health and Human Services, Centers for Disease Control and Prevention, National Center for Environmental Health. 2005. Third national report on human exposure to environmental chemicals. http://cdc.gov/exposurereport/pdf/thirdreport.pdf
World Wildlife Fund. 2005. Generations X: Results of World Wildlife Fund's European Family Biomonitoring Survey. http://assets.panda.org/downloads/generationsx.pdf

SUGGESTED READING

Hill, A. B. 1965. The environment and disease: Association or causation? *Proc. Roy. Soc. Med.* 58 (5): 295–300.

5 The Cornerstone of Risk Assessment
Toxicity Testing in Animals

5.1 INTRODUCTION

Protection of public health from toxic chemical risk is only occasionally based on human data, for two reasons. First, the number of chemicals for which direct human toxicity data are available is small, because it is limited to unintentional exposures. Second, even when available, human epidemiological studies generally detect the effect of a toxic chemical on 1 in 100 (10^{-2}) to 1 in 1,000 (10^{-3}) people (Figure 4.1). That is not sensitive enough to serve as a basis for protecting public health. Regulating a chemical to a risk of 1 in 1,000 would mean that in a city of 1 million people, about 1,000 people could get sick.

To safeguard public health, methodologies are needed that are more sensitive than human epidemiological studies and that can be applied to any and all chemicals of concern. The most useful methodology for assessing human health risk is toxicity testing in small mammals. Basing human health-risk assessment on animal data is biologically plausible, because we are closely related to other mammals in evolutionary terms, and we generally respond similarly, although not identically, to toxic chemicals. Given our biological similarities, toxicity data in small mammals can be extrapolated to humans with a reasonable degree of confidence and used as a basis for human health-risk assessments (Chapter 8).

A number of federal laws are intended to protect the public from various classes of hazardous chemicals. Implementation of these laws depends primarily on data from toxicity testing in animals (Table 4.1). Federal agencies that administer laws regulating toxic chemicals include the U.S. Environmental Protection Agency (EPA), the Food and Drug Administration (FDA), and the Occupational Safety and Health Administration (OSHA). Examples of chemical compounds that may be subjected to toxicity testing include new drugs, food additives, cosmetics, and pesticides. Federal agencies perform only a limited amount of toxicity testing themselves. Instead, corporations pay for toxicity testing as part of the cost of bringing a chemical to market. Researchers at universities may also perform toxicity testing with grant support from corporations or government agencies. By and large, the role of government is limited to reviewing the results of toxicity tests performed by others. In addition to new products, an agency may reevaluate a chemical or a drug that is already on the market but that new evidence suggests may be more toxic than originally believed.

TABLE 5.1
Organisms Commonly Used in Toxicity Testing

Type of Organism	Organism
Invertebrates	*Daphnia magna*
	Crayfish
	Mayflies
	Midges
	Planaria
Aquatic vertebrates	Rainbow trout
	Goldfish
	Fathead minnow
	Catfish
Algae	*Chlamydomonas reinhardtii* (green algae)
	Microcystis aeruginosa (blue-green algae)
Mammals	Rats
	Mice
Avian species	Bobwhite
	Ring-necked pheasant

Source: Reprinted with permission from Ian Pepper, Charles Gerba and Mark Brusseau, eds., *Pollution Science* (San Diego: Academic Press, 1996), 329.

EPA is also charged with regulating the release into the environment of the chemical byproducts of industrial processes, for example, diesel exhaust and mercury.

Most toxicity testing aimed at protecting human health is performed in rats, mice, guinea pigs, and rabbits. In addition to their evolutionary proximity to humans, they offer the practical advantage of being small and therefore less expensive than larger animals to maintain in a laboratory setting. Nevertheless, toxicity testing even in small animals is quite expensive. A typical test costs thousands to hundreds of thousands of dollars. A carcinogenicity (cancer) test in rats costs about $1.4 million.

A number of federal laws protect the environment as well as human health. The discipline of ecotoxicology encompasses toxicity testing in hundreds of nonmammalian species that are evolutionarily distant from our own, for example, fish, algae, zooplankton, higher plants, and birds (Table 5.1). Data obtained in ecotoxicological studies support ecological risk assessments (Chapter 9).

5.2 DESIGNING A TOXICITY TEST

When designing a toxicity test, there are four primary considerations:

1. The route of exposure, i.e., oral, inhalation, or dermal
2. The magnitude of the doses in the test
3. The time frame of exposure, usually either acute or chronic (explained below)

The Cornerstone of Risk Assessment 69

4. The specific toxic effect of concern, or endpoint, to be studied in the test, e.g., death, infertility, or cancer

In addition, statistical power and cost are important factors in test design.

5.2.1 ROUTE OF EXPOSURE

The three routes by which humans are exposed to toxic chemicals are oral, i.e., eating food to which chemicals are adsorbed and drinking water in which chemicals are dissolved; dermal, i.e., absorbing a chemical through the skin; and inhalation, i.e., breathing in chemical vapors or dust particles. Chemicals entering the body through the digestive tract, the skin, or the lungs may be taken up into the bloodstream and transported throughout the body by the circulatory system. The entry of a chemical into the bloodstream is generally required for a toxic effect to occur. However, a few chemicals exert their toxic effects locally. For example, inhaled asbestos and particulate matter act directly on the lungs. The organs and physiological processes responsible for the body's uptake, distribution, biotransformation, and elimination of chemicals are introduced in Chapter 6.

Toxicologists distinguish the route of exposure from the pathway of exposure. The route of exposure is the portal—eating, breathing, or touching—by which a chemical enters the body of a biological receptor, the term toxicologists use to refer to a human being or other exposed organism. The pathway of exposure, on the other hand, is the environmental itinerary a chemical travels as a result of partitioning into, and being advectively transported by, one or more of the three environmental media—water, air, and soil/sediment—to the point of contact with the biological receptor (Chapter 2). The route of exposure is a critical component in the design of every toxicity test because the dose required to elicit a specific toxic effect, or indeed the toxic effect itself, often depends on it. The route of exposure is a critical part of the toxicity test design and is always specified when reporting test results.

5.2.2 DOSE AND TIME FRAME

To approximate real-world exposure scenarios, the time frame of toxicity testing is generally divided into two broad categories: acute and chronic. Acute toxicity typically involves exposure to a single large dose, and the toxic effect of concern is manifested within hours to days. The outcome of an acute-toxicity test in small mammals is defined as the number of animals manifesting the toxic effect of concern—for example, death—within 14 days following administration of a single dose of the test chemical. Chronic toxicity is defined as a toxic effect that results from exposure to repeated small doses of a chemical over a period of years. A chronic toxicity test in small mammals is defined as lasting six months or longer; it may cover the entire adult life span of the test species, two years in mice and two-and-a-half years in rats. Tests to study the carcinogenicity (cancer-causing potential) of a chemical are chronic toxicity tests designed to last the entire adult lives of the test animals. Subchronic toxicity is sometimes used to describe a toxic effect that results from repeated exposure over a time period that is intermediate between acute and chronic toxicity, i.e., weeks to months.

Reproductive toxicity is defined as an adverse effect on the ability of males or females to produce viable offspring. The time frame of reproductive toxicity testing is determined by the reproductive cycle of the test species and involves one or more generations of offspring. Single-generation tests involve a single cycle of breeding, gestation, and birth. In multigeneration tests, the parents, the first generation of offspring, and the second generation of offspring (referred to as the F0, F1, and F2 generations, respectively) are exposed and monitored for adverse effects on fertility.

5.2.3 Endpoint or Specified Toxic Effect

The endpoint that is chosen for a toxicity test must fulfill two criteria: It must be biologically relevant to the risk the test seeks to assess, and it must be susceptible to accurate measurement. Death is the quintessential example of an endpoint that is both biologically relevant and readily quantifiable. Other examples are infertility and cancer. Common to such clear-cut endpoints is the biological either-or quality of the toxic effect. However, many toxic effects involve gradations, for example, anemia, loss of hearing, and low birth weight. When toxicity involves a graded effect, a cutoff must be specified as the endpoint, for example, a 35% reduction in red blood cells, 50% hearing loss, or 20% reduction in birth weight.

A chemical may have a measurable effect, but the biological significance of the effect may be in doubt. A measurable effect that may or may not be indicative of disease is called a biomarker. For example, epidemiological evidence may suggest that exposure to a particular chemical is associated with a statistically significant decrease in sperm count, but fertility is not affected. Another chemical might be associated with a small increase in the activity of a liver enzyme but not with an increased incidence of liver disease. Should any measurable change in a biological system, regardless how small, be considered indicative of toxic chemical risk? This question is particularly relevant to risks that may be associated with long-term, low-dose exposure. Toxic effects and biomarkers of exposure may or may not be related. If they are related, the next question is whether there are biomarker thresholds that signal the onset of toxic effects. Biomarkers are rarely used as endpoints in toxicity tests unless their toxicological significance is well understood.

5.2.4 Statistical Power and the Cost of Toxicity Tests

The statistical power of human epidemiological studies is increased by increasing the sizes of the samples drawn from the at-risk and control populations (Chapter 4). Animal toxicity tests are, in effect, epidemiological studies of animal populations under controlled laboratory conditions. As in human studies, sample size directly affects the statistical power of the toxicity test. However, using thousands of animals is impractical because of the expense involved in housing and caring for them in the laboratory. Instead of using large numbers of animals, statistical power is increased by raising the dose of the chemical in order to increase the frequency of the specified toxic effect. The toxic effect frequency can be estimated from small pilot studies,

5.3 DESCRIPTIONS OF TOXICITY TESTS AND THEIR PRODUCTS

5.3.1 Acute Lethality

Probably the most widely performed toxicity test, acute lethality is often investigated in rats. The quantal toxic effect of concern, or endpoint, is death. Typically, an acute lethality test consists of approximately five dosage groups of 10 to 20 animals each plus a control group. A single dose of the test chemical is administered at the beginning of the study. The animals are examined daily, and clinical signs and symptoms of toxicity are recorded. After an interval of 14 days in which no further exposure takes place, the numbers of dead animals in each dosage group and the control group are counted, and the results are analyzed statistically with respect to the excess death rate among the exposed animals as a function of dose. Figure 3.1 presents an overview of the general statistical approach. In the context of an acute lethality study, the median dose is referred to as the median lethal dose, or LD_{50}. If the endpoint is an effect other than death, the median dose is referred to as the median toxic dose, or TD_{50}. It is important to appreciate that the median dose is a statistic, not an actual dose. In the case of the median lethal dose, it is the statistically derived dose above which half of the test animals die and below which half survive. When reporting the LD_{50}, both the test species and the route of exposure are given, because both are major factors determining the LD_{50}. For example, the LD_{50} for arsenic in rats exposed by oral ingestion is reported as follows: $LD_{50\ oral,rat} = 48$ mg/kg. Oral LD_{50} values in rats are listed in Table 5.2 for several familiar chemicals. A separate toxicity test must be performed for each of the three routes of exposure that is anticipated for human exposure to the chemical: ingestion, inhalation, and/or dermal absorption.

The LD_{50} in rats is a useful statistic because it provides a general indication of a chemical's toxicity. It is not the same as an LD_{50} in humans, however. The LD_{50} in humans is and should be unknowable, of course, because moral and ethical considerations prevent experimentation on humans. The human LD_{50} might be estimated from accidental poisoning incidents, but it cannot be determined accurately, generally because the degrees of exposure vary uncontrollably. LD_{50} values in rats are relevant to human health-risk assessment in two ways. First, because of biological similarities between rats and humans, rat LD_{50} values usually provide ballpark estimates of human LD_{50} values. In other words, a chemical that is very toxic in rats is also likely to be very toxic in humans. But evolutionary relatedness does not guarantee that toxic chemicals have similar potencies in different species. Toxic doses can and do vary significantly across species lines. Depending on the chemical compound and route of exposure, the doses at which symptoms of toxicity appear in rats and humans—as expressed in units of milligrams of chemical per kilogram of body weight—may differ by a factor of 10, 100, or more.

TABLE 5.2
Approximate Acute Oral LD_{50} Values of Selected Chemicals in Rats

Agent	LD_{50} (mg/kg)
Sugar	29,700
Polybrominated biphenyls (PBBs)	21,500
Alcohol	14,000
Methoxychlor	5,000
Vinegar	3,310
Salt	3,000
Malathion	1,200
Aspirin	1,000
Lindane (benzene hexachloride delta isomer)	1,000
2,4-D	375
Ammonia	350
DDT	100
Heptachlor	90
Arsenic	48
Dieldrin	40
Strychnine	2
Nicotine	1
Dioxin (TCDD)	0.001
Botulinus toxin	0.00001

Source: Reprinted with permission from Michael Kamrin, *Toxicology. A Primer on Toxicology, Principles and Application* (Chelsea, MI: Lewis Publishers, 1988), 46.

Note: Comparison of LD_{50} values provides a basis for assessing the relative potencies of toxic chemicals.

While the absolute toxicity of a chemical can vary by orders of magnitude from species to species, the relative toxicities of chemicals tend to be more conserved through evolution. Thus if chemical B is more toxic than chemical A in rats, and A is more toxic than C, the same order of toxicity, B > A > C, is likely to hold in humans, as well. Because of the greater evolutionary conservation of relative toxicities, LD_{50} values in rats in effect provide an approximate scale for ballparking the relative toxicities of chemicals in humans. For example, strychnine is 50 times more lethal than DDT in rats (Table 5.2), and therefore it is reasonable to infer that strychnine is probably on the order of 10 to 100 times more lethal than DDT in humans. The availability of scales of relative toxicities in a few species, such as rats, fathead minnows, ring-necked pheasants, and water fleas, are valuable for inferring relative toxicities in related species.

In addition to its usefulness as a rough scale of chemicals' relative potencies as lethal agents, acute-toxicity testing also provides preliminary information on two other important characteristics of toxicity. Information about the reversibility of toxicity can be obtained by counting the number of animals in each dosage group that, instead of dying, recover spontaneously from acute exposure. In addition, the mechanism of toxicity can be investigated by performing autopsies on dead animals and noting which internal organs have been affected by exposure to the chemical.

In summary, acute-toxicity testing in small mammals such as rats provides a range of valuable information on chemical toxicity, including lethal doses by different routes of exposure, mechanism(s) of lethal toxicity, the extent to which the lethal effect may be reversible in some members of the exposed population while others die, and the relative potencies of toxic chemicals as lethal agents.

5.3.2 Subchronic Toxicity Testing

Unlike the one-time exposure in an acute lethality test, a subchronic toxicity test involves repeated doses of a test chemical, typically administered over a period of approximately 90 days. The goals of a subchronic toxicity test are to investigate organ toxicity and obtain dose–effect data with which to design a chronic toxicity test, including an estimate of a "no observed adverse effect level," or NOAEL (see Section 5.3.3). At least three dose levels are tested: a high dose chosen to cause 10% or fewer mortalities; a low dose chosen to produce no toxic effects; one or more intermediate doses; and a control group that is not exposed to the test chemical. Two species may be tested, for example, rats and dogs. In addition, separate tests are generally performed on males and females, because gender can affect the way the body responds to a toxic chemical. Thus, as many as four subchronic toxicity tests may be conducted per chemical for each predicted route of exposure.

Subchronic toxicity studies do not seek to determine an LD_{50} or other statistical parameters; hence statistical power is less important in a subchronic toxicity test than in acute and chronic toxicity tests. Dosage groups contain relatively small numbers of animals, e.g., 10 to 20 rats and 4 to 6 dogs. Exposed animals are observed closely for signs and symptoms of toxicity. Blood samples are collected and analyzed at regular intervals. At the end of 90 days, all surviving animals are sacrificed and autopsied, including microscopic examination of organs and tissues to characterize the pathologies associated with exposure to the test chemical.

5.3.3 Chronic Toxicity and Carcinogenicity Testing

In a chronic toxicity test, repeated exposures are continued for six months or longer. The duration of a chronic toxicity test depends on the anticipated duration of human exposure relative to the human life cycle. For example, if the chemical is a food additive to which humans may be exposed throughout their adult life, the chronic toxicity test is designed to last the adult life span of the test species, about two years in mice and two-and-a-half years in rats.

Whenever feasible, chronic toxicity tests are designed to include both cancer and other long-term toxic effects. A test of the carcinogenicity of a chemical is, by definition, a chronic toxicity test, because cancers typically show a lag time of years, and sometimes decades, between exposure and the onset of disease. Because of their biological differences, females and males are again tested separately in a chronic toxicity test. Three dosage groups are typically employed for each sex: high, intermediate, and low, plus an unexposed control group. Statistical power is an important consideration, and a dosage group typically contains about 60 animals for a total of approximately 240 animals per sex per test. The highest dose, referred to as the maximum tolerated dose (MTD), is selected on the basis of the results of subchronic toxicity studies. The choice of the MTD is designed to result in the deaths of not more than half of the test animals, as mortality greater than 50% would reduce the statistical power of the study. The intermediate and low doses are typically set equal to 1/2 and 1/4 of the MTD, respectively. When cancer is the endpoint of the chronic toxicity test, the MTD is defined as the highest dose resulting in death by cancer only. That is, doses higher than the MTD cause death as a result of cancer as well as other toxicities, while doses equal to and lower than the MTD kill only by causing cancer.

The test animals are monitored closely for signs of toxicity, and blood samples are collected regularly. Autopsies are performed on animals that die in the course of the study. Animals that survive the study are sacrificed and autopsied. Chronic toxicity testing provides further insight into the pathology(ies) and mechanism(s) of disease caused by exposure to a toxic chemical. It establishes a relationship between the degree of exposure to a chemical and the frequency with which a toxic endpoint such as cancer is manifested in the exposed population.

5.3.4 Reproductive Toxicity Testing

Reproductive toxicity refers to the adverse health effects in the offspring of parents exposed to a toxic chemical. In mammalian toxicology, two broad classes of reproductive toxicity tests are performed: single generation and multigeneration, usually in rats. Single-generation toxicity testing is divided into three subclasses referred to as Segments I, II, and III. The segments differ with respect to the timing and duration of exposure and the toxicity endpoints that are scored. For example, Segment I, the most inclusive of the single-generation toxicity test protocols, exposes parents beginning several weeks prior to mating to include possible effects of the chemical on spermatogenesis, estrous, and mating behavior, and it continues through gestation (pregnancy) and parturition (birth) until weaning. Endpoints in Segment I toxicity tests include mating behavior, gonadal function, and fertility as well as maternal behavior and the development and growth of the offspring as embryo, fetus, and newborn. Animals are sacrificed and autopsied to investigate toxicity to organs, particularly the gonads.

Segment II of the single-generation test focuses on the teratogenic potential of the test chemical (ability to cause birth defects). Teratogenicity is also referred to as developmental toxicity. The Segment II protocol calls for exposing the embryo by administering the chemical to pregnant females between days 6 and 15 of gestation. This period corresponds to the first trimester of a human pregnancy, a time when the embryo's organs are developing and it is most vulnerable to chemical injury. Typical toxic effect endpoints are the viability, weight, and gender of the offspring and the presence of birth defects, both gross defects (visible to the naked eye) and microscopic defects that are detected when the newborn animals are sacrificed and autopsied.

The Segment III protocol of single-generation reproductive toxicity testing exposes the mother during the perinatal and postnatal periods, beginning with day 15 of gestation and continuing until weaning. Toxic effect endpoints include the growth of the fetus and the newborn, maternal behavior, and weaning success. Taken together, the results of Segments I, II, and III profile a chemical's effects on the reproductive health of a species due to exposure at any point in the reproductive cycle, from the eggs and sperm of the parents through the birth of the offspring and up until it reaches sexual maturity at puberty.

Multigeneration reproductive toxicity testing investigates the transgenerational toxicity of a chemical, particularly including genetic effects that may take more than one generation to be expressed. Testing is typically performed across three generations of animals: the parental (F0), F1, and F2 generations. Each generation is subjected to continuous exposure, and each is scored for reproductive toxicity. Toxic effect endpoints are similar to those in Segment I of the single-generation test. Statistical significance is a factor in reproductive toxicity testing, and the doses and numbers of animals per dosage group are chosen to maximize statistical power.

5.3.5 Toxicity Test Design in Nonmammalian Species

The general design of toxicity tests in nonmammalian species is similar to test design in mammalian species. Differences are due mainly to biology such that exposure conditions are designed to simulate the test species' contact with the chemical in their natural habitat. For example, fish may be exposed by dissolving a test chemical in water, earthworms by allowing a test chemical to adsorb to soil, and benthic organisms by adding a test chemical to sediment.

As one example of tests on nonmammalian species, aquatic toxicity testing on fathead minnows is divided into two types, acute and chronic. In an acute aquatic toxicity test, lethality is usually the endpoint of choice. The test chemical is commonly administered by dissolving it in water. The concentration of the chemical in water is different for each dosage group, and the control group of fathead minnows is maintained in water free of test chemical. The duration of exposure is 24, 48, or 96 hours. The major output of the test is the median lethal concentration, or LC_{50}, a statistic that is defined analogously to the LD_{50} as the dissolved concentration of test chemical that produces a cumulative mortality of 50% of the test population (Figure 3.1b).

Chronic toxicity testing seeks to evaluate the ability of a population to maintain or increase its size when subjected to long-term exposure to sublethal concentrations of a chemical. A population will decline and die out over a period of years if reproductive rates fall below a certain level. Chronic toxicity tests may be full life cycle or partial life cycle. A full life-cycle test examines the effect on fertility of exposing organisms to a toxic chemical throughout their life cycle. The duration of a full life-cycle test is determined by the biology of the test species. It is 30 days for the fathead minnow, and for *Daphnia magna*, a water flea, it is 21 days, to give two examples. A partial life-cycle test investigates one (or sometimes more than one) sensitive stage in the life cycle of a species. The timing and duration of exposure correspond to the time the species spends in its sensitive life stage(s). A major product of a chronic toxicity test, either full life cycle or partial life cycle, is the maximum acceptable toxicant concentration (MATC). The MATC represents an attempt to quantify a threshold of exposure below which a population can maintain its size and above which the decline and extinction of the population are likely.

5.4 THE PROBIT PLOT

Toxicity tests are, in effect, small-scale epidemiological studies under controlled laboratory conditions of the incidence of chemically induced disease as a function of exposure. As outlined in Chapter 3, there are two basic statistical approaches for analyzing the results of toxicity tests: a histogram that relates dose to the *incremental* incidence of the specified toxic endpoint (Figure 3.1a), and a sigmoidal curve that relates dose to the *cumulative* incidence of the toxic endpoint (Figure 3.1b). For purposes of data analysis, the cumulative dose–effect curve is more advantageous than the incremental dose–effect histogram because it greatly facilitates determination of a valuable statistical parameter, the median dose. However, the nonlinearity of the cumulative dose–effect curve may undermine its effectiveness as an analytical tool in some circumstances. Toxicity testing is always constrained by cost, and there are typically only three or four dosage groups in chronic toxicity tests, e.g., tests for carcinogenicity. The smaller the number of data points, the more problematic it is to characterize the mathematical relationship between dose and effect using the sigmoidel cumulative dose–effect curve. This, in turn, leads to another difficulty: The poorer the mathematical relationship between exposure and effect, the more difficult it is to extrapolate from the high doses required for statistical significance in a laboratory study of chronic toxicity to the low doses characteristic of most real-world human encounters with toxic chemicals.

The probit plot is a straight-line plot that facilitates curve-fitting. It eases but does not eliminate the difficulties of working with limited dose–effect data. The probit plot is compared with the two other tools for analyzing dose–effect data, the histogram and the sigmoidal curve, in Figure 5.1. Like the sigmoidal curve, the probit plot tracks cumulative, not incremental, toxic endpoint frequency as a function of increasing exposure. Further, the probit plot, like the sigmoidal curve, is based on the assumption that the incremental frequency of a specified toxic effect is distributed normally, i.e., that the frequency distribution is represented (under ideal conditions) by a bell-shaped curve.

Figure 5.1 shows that all three graphs used to characterize the relationship between the dose of toxic chemical and the frequency of toxic effect—i.e., the frequency histogram, the cumulative dose–effect curve, and the linear probit plot—put the logarithm of the dose on the x-axis. The reason the probit plot is linear is because of a clever invention: a statistically derived scale that represents cumulative-effect frequency on the y-axis. Called probit units, or simply probits, this scale is based on a particular statistic, the standard deviation of the mean. Standard deviations correspond to fixed percentages of a population, and they can therefore be used in place of percentages to represent the fraction of a population that manifests a toxic effect. The mean dose corresponds to a standard deviation of zero because it is located in the exact middle of the bell curve. In terms of standard deviations, a disease frequency of 50% of the population is equivalent to zero standard deviations. One standard deviation below and above the mean corresponds, respectively, to manifestation

The Cornerstone of Risk Assessment

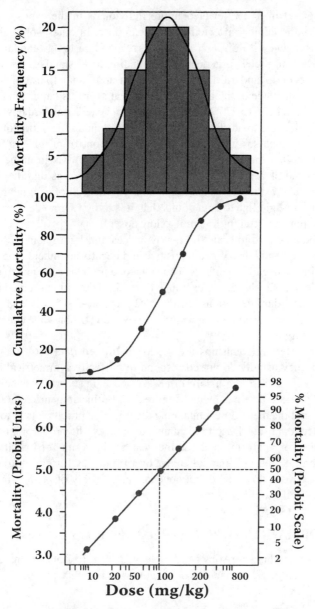

FIGURE 5.1 Three representations of the increase in the frequency of a toxic effect (death) as a function of toxic chemical dose. Top: The frequency of mortality at each dose as a percent of the population, also referred to as the incremental frequency of the effect (similar to Figure 3.1a). Middle: Total mortality at each dose as a percent of the population, also referred to as the cumulative frequency (similar to Figure 3.1b). Bottom: Same as middle panel: Total mortality at each dose as a percent of the population, also referred to as the cumulative frequency, but with mortality scaled in probit units instead of percent. Note the use of a logarithmic scale for dose on the x-axis. (Reprinted with permission from Curtis Klaassen, ed., *Casarett & Doull's Toxicology: The Basic Science of Poisons* [New York: McGraw-Hill, 2001], 19.)

of the toxic effect in 34.1% fewer or more individuals in the exposed population compared with the number affected at the mean dose. In other words, the cumulative disease frequency is 15.9% at a dose corresponding to −1 standard deviation below the mean and 84.1% for a dose corresponding to +1 standard deviation above the mean. A second standard deviation spans an additional 13.6% of the exposed population on either side of the mean, meaning that over this larger range of chemical exposure, from 2.3% to 97.7% of the population, i.e., ±2 standard deviations from the mean, manifest the toxic effect. Adding a third standard deviation on either side of the mean encompasses almost the entire population, from 0.1% to 99.9%. The advantage of scaling the toxic effect frequency on the y-axis in units of standard deviations instead of percentages of the exposed population is that the dose–effect relationship can be expressed as a straight-line plot instead of a sigmoidal curve. It is easier to fit a straight line to data points than it is to fit a sigmoidal curve. It is also easier to extrapolate a straight-line plot to low doses.

It would be awkward to scale the y-axis in plus and minus standard deviations, because the scale would then contain positive and negative numbers. The probit convention eliminates negative numbers. The value of a probit is set equal to the number of standard deviations plus 5. For example, 0 standard deviation is equal to 5 probits (0 + 5 = 5), −3 standard deviations is equal to 2 probits (−3 + 5 = 2), and +1 standard deviation is equal to 6 probits (+1 + 5 = 6). The resulting probit scale consists of positive numbers only.

Most dose–effect relationships are based on limited data sets, and when they are analyzed, they usually do not correspond to perfectly symmetrical bell-shaped curves. Dose–effect histograms are often skewed, and in practice the mean dose and the median dose rarely coincide (Figure 3.1). Probit units are referenced to the median dose, not the mean dose, enhancing the utility of probit plots as tools for analyzing nonideal data sets. Despite its many advantages, the probit plot is not always the best choice for analyzing and communicating the results of toxicity tests. The choice of a probit plot, a sigmoidal curve, or a frequency histogram to portray the relationship between dose and effect depends on a variety of factors, including the audience who will use the results.

5.5 INFORMATION DERIVED FROM TOXICITY TESTING

Several important characteristics can be gleaned from toxicity testing and used to assess human health risk (Chapter 8) and ecological risk (Chapter 9).

5.5.1 TOXIC EFFECT FREQUENCIES RESULTING FROM SPECIFIC EXPOSURE LEVELS

The most common statistic obtained from animal toxicity testing is the median dose. To have meaning, the median dose needs to be reported in the context of the toxicity test from which it is derived. For example, if the toxicity test was for acute lethality, then the median dose is reported as the 50% lethal dose, or LD_{50}; the species and route of exposure are also specified, e.g., $LD_{50\,oral,rat}$. If it is a long-term or chronic toxicity test with an endpoint other than death, e.g., liver disease, the median dose is reported as the 50% toxic dose, or $TD_{50\,oral,rat}$.

The quantal dose–effect curve provides a guide to toxic effect frequencies relative to exposure under the conditions of the laboratory toxicity test. Doses corresponding to other toxic effect frequencies may be extrapolated from cumulative dose–effect curves. For example, the 10% toxic dose, or TD_{10}, represents a dose corresponding to a cumulative-effect incidence of 10% of the test population; it is determined by finding the dose that corresponds to a 10% cumulative-effect frequency (in the sigmoidal curve) or 3.7 probit units (in the probit plot). The 75% toxic dose, or TD_{75}, is the dose corresponding to 75% cumulative-effect incidence or 5.8 probits.

5.5.2 Threshold of Toxicity

For many chemicals, available evidence as well as biological theory suggest that a threshold level of exposure exists below which the chemical does not cause harm to the organism, because the body's defenses are able to ward off the chemical assault (Chapter 6). Threshold doses cannot be measured directly, because by definition a threshold is a dose above which a toxic effect first starts to become measurable. That is, any toxic effect that can be measured is, by definition, above the threshold of toxicity. Thresholds may be extrapolated from cumulative dose–effect curves; however, there is generally more than one reasonable model for performing an extrapolation below the range of experimental data, and therefore the true threshold dose is difficult to determine with confidence (Figure 5.2).

A toxicity threshold is typically estimated from a short-term toxicity study on a single chemical with a clearly defined toxic endpoint. Estimating a threshold dose—indeed, investigating whether a threshold exists—is more difficult when exposure is to low doses of a chemical over long periods of time. As a rule, the number of animals required to identify a toxicity threshold in long-term, low-dose studies is prohibitively expensive, and such studies are rarely, if ever, performed. The question of toxicity thresholds is complicated further because real-world exposures are typically not to one chemical, but to multiple chemicals in changing combinations over time. More studies are needed to characterize thresholds of toxicity both for single chemicals and chemical mixtures.

Low-dose exposure is of interest for other reasons besides toxicity thresholds. Some chemicals are beneficial at low doses, for example, essential nutrients such as vitamins. The dose–effect curves for these chemicals are U-shaped: Too little exposure is harmful because it results in a deficiency disease, and too much exposure causes toxic effects; the toxic effects are unrelated to the harmful effects of deficiency (Figure 5.3). It has been suggested, although not yet proven, that radiation as well as alcohol and several other chemicals provide beneficial effects at low doses, even though they are harmful at high doses, a concept referred to as hormesis. On the other hand, there are also reports of chemicals exhibiting greater toxicity at low doses than at higher doses. If true, such low-dose phenomena would suggest that the dose–effect relationship may not apply to some chemicals. Possible exceptions to the dose–effect relationship deserve further investigation.

From a practical standpoint, the "no observed adverse effect level" (NOAEL) offers a facsimile of the threshold that is measurable and that can be used in assessing and managing risk (Point E in Figure 5.2). The NOAEL is defined as the dose

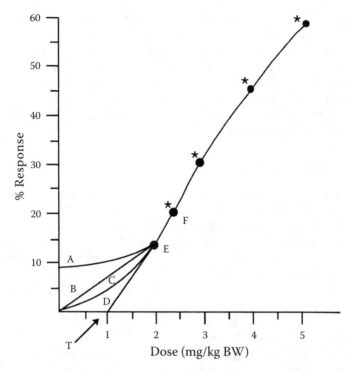

FIGURE 5.2 Lowest observed adverse effect level (LOAEL), no observed adverse effect level (NOAEL), and threshold. The figure portrays a typical dose–effect (also referred to as dose-response) curve based on five dosage groups. The x-axis is scaled in units of milligrams of toxic chemical per kilogram of body weight (mg/kg BW). The starred points (*) indicate dosage groups manifesting the toxic endpoint at a statistically significant frequency. The LOAEL is defined as F, the lowest dose with a statistically significant frequency of the toxic effect among the 30 or so animals in the dosage group. The NOAEL is defined as E, the lowest dose at which the toxic effect is unmistakably observed, but at too low a frequency to achieve statistical significance. Estimates of the threshold of toxicity depend on the particular model that is used to extrapolate to doses below E, where no experimental data exist. Model A predicts there is no threshold at all; rather, it predicts a background toxic effect frequency of 9%. Models B and C predict a threshold of zero, i.e., any and all increases in dose above zero produce the toxic effect, albeit at different frequencies in the two models. Model D predicts a threshold of 1 mg/kg body weight, meaning that at doses from 0 to 1 mg/kg, the body is able to defend itself against the chemical and there is no toxic effect. (Reprinted with permission from Curtis Klaassen, ed., *Casarett & Doull's Toxicology: The Basic Science of Poisons* [New York: McGraw-Hill, 2001], 92.)

in a toxicity study that produces the toxic endpoint in a few animals but not enough to be statistically significant. In other words, the NOAEL is the lowest dose causing a toxic effect that is observable, but the number of animals that is affected is too small to perform meaningful statistical calculations. It is estimated that 5% to 10% of an exposed population, presumably the most sensitive individuals, are affected at the NOAEL. Thus, the NOAEL is not a true threshold. It is useful because it offers a measurable basis for assessing human health risk (Chapter 8). In addition to the

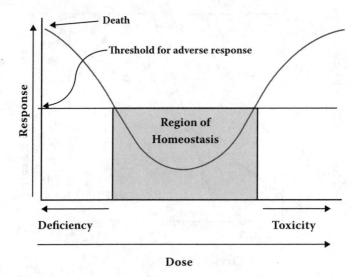

FIGURE 5.3 Dose–effect curve for essential chemicals. For a few chemicals such as vitamins or trace elements, deficiencies as well as high doses can result in adverse effects, including death. The symptoms of the deficiency syndrome and the toxic effects syndrome are different. The range of doses that sustain life is referred to as the region of homeostasis. (Reprinted with permission from Curtis Klaassen, ed., *Casarett & Doull's Toxicology: The Basic Science of Poisons* [New York: McGraw-Hill, 2001], 21.)

NOAEL, a toxicity study also seeks to identify the "lowest observed adverse effect level" (LOAEL), defined as the lowest dose producing a biologically significant toxic effect in a statistically significant number of animals (Point F in Figure 5.2).

In summary, the NOAEL has biological significance but not statistical significance, whereas the LOAEL has both biological and statistical significance. The NOAEL and the LOAEL are important factors in human health-risk assessment.

5.5.3 The Rate of Increase in Chemical Disease Frequency as a Function of Dose

The steepness, or slope, of a cumulative dose–effect curve reflects the range of exposure levels over which the toxic endpoint is manifested (Figure 5.4). The steeper the slope, the smaller is the dose range producing the toxic effect and the larger is the increase in toxic effect frequency per unit dose. The shallower the slope, the larger is the dose range and the smaller is the increase in toxic effect frequency per unit dose. The reasons chemicals vary with respect to their dose ranges are complex and may involve multiple factors. One major factor is the nature of the chemicals themselves. Genetic variability within the test population probably also has a role. For example, cumulative dose–effect curves tend to be somewhat steeper for colonies of inbred laboratory rats than for feral rats, probably because populations of wild animals have a bigger gene pool. To the extent possible, scientists use genetically similar strains of test animals to compare the toxic effects of different chemicals. The steepness of the slopes of cumulative dose–effect curves (Figure 5.4) is of considerable significance

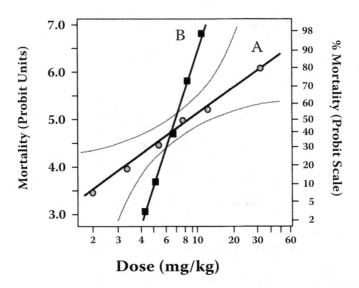

FIGURE 5.4 Slope of the log-dose effect curve. Two different chemicals may produce the same toxic effect but at different frequencies due to differences in the molecular structures of the chemicals. If there are significant biological differences between two at-risk populations, the same chemical may produce its toxic effect at different frequencies. Toxic effect frequency A is presented together with its 95% confidence limits. Toxic effect frequency B is steeper than A. The steeper slope translates into a greater increase in the frequency of death or other toxic effect per unit increase in exposure. Note that toxic effect frequency B may correspond to a different chemical and the same exposed population as A; alternatively, toxic effect frequency B may correspond to the same chemical as A but a more vulnerable at-risk population. The slope of the dose–effect curve has ramifications for protecting public health. (Reprinted with permission from Curtis Klaassen, ed., *Casarett & Doull's Toxicology: The Basic Science of Poisons* [New York: McGraw-Hill, 2001], 20.)

to public health. A steeper slope translates into smaller increases in exposure resulting in higher percentages of the exposed population experiencing the toxic effect. The slope of the cumulative dose–effect curve is used extensively in cancer risk assessment (Chapter 8).

5.5.4 Potency and Efficacy

These two terms come from the field of medical pharmacology, and although they are used to describe a chemical's toxicity, they are more meaningful in the context of pharmacological therapeutics. Potency is the term that is used to refer to the median dose eliciting a drug's therapeutic effect; when referring to drugs, the median dose is called the 50% effective dose, or ED_{50}. In other words, the ED_{50} of a drug and the potency of a drug are synonymous terms. Therefore, if two chemicals, A and B, produce the same therapeutic effect, but the ED_{50} for chemical B is lower than the ED_{50} for chemical A, then chemical B is said to be more potent than chemical A with respect to that specific therapeutic effect (Figure 5.5). In the context of toxic chemicals, lower LD_{50} or TD_{50} values are also said to indicate greater potency. For

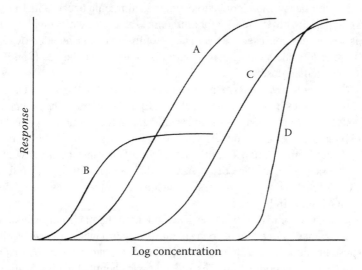

FIGURE 5.5 Potency, efficacy, and toxicity. The terms *potency* and *efficacy* are used to characterize dose–effect curves for beneficial chemicals, e.g., drugs, as well as dose–effect curves for toxic chemicals. They apply both to graded dose–effect curves on individual test animals and to cumulative dose–effect curves on populations of test animals. In a graded dose–effect curve, potency refers to the dose that elicits 50% of the maximum physiological effect in an individual organism (therapeutic in the case of a beneficial drug; toxic in the case of a harmful chemical) (see Figure 3.2). In a cumulative dose–effect curve, potency refers to the dose at which a chemical elicits a specific effect, either therapeutic or toxic, in 50% of the population. In other words, in the context of a population study, potency is equivalent to the median dose: the lower the median dose, the more potent the chemical (Figure 3.1b; Figure 5.1, bottom). The sigmoidal curves in Figure 5.5 can be interpreted as a set of graded dose–effect curves on an individual animal (that is how it is described in the original source); alternatively, Figure 5.5 can also be interpreted as a set of cumulative dose–effect curves on a population of animals. The order of potency, from highest to lowest, is B > A > C > D. Efficacy refers to the maximum intensity of a chemical's physiological effect in the context of a graded dose–effect curve; in the context of a cumulative dose-response curve, efficacy refers to the maximum percentage of a population showing a specific response. (Note: The terms *dose–effect* and *dose-response* are synonymous.) When interpreting Figure 5.5 from a population perspective, chemicals A, C, and D produce the desired response in a similar percentage of individuals and therefore have similar efficacies. Fewer individuals show the desired response with the less efficacious drug B. Note that it is possible for a chemical (such as B) to be more potent, yet less efficacious than other chemicals that produce the same effect. While the term *toxicity* is properly used to describe one of several classes of harmful effects that chemicals possess (Chapter 1), it is often used to refer to the potency of a toxic chemical (e.g., see Table 5.2). To avoid confusion, it is suggested that the term *potency* be reserved for the median dose (in a population study) as well as the 50% effective concentration (in an individual animal) and that *toxicity* be used to refer to the general class of harmful effects at the cellular and molecular level. (Reprinted with permission from Bertram G. Katzung, *Basic and Clinical Pharmacology*, 5th ed. [Norwalk: Appleton & Lange, 1992], 28.)

example, strychnine is a more potent toxicant than ammonia because its $LD_{50\,oral,rat}$ is 2 mg/kg compared with 350 mg/kg for ammonia (Table 5.2). Unfortunately, the term *toxicity* is sometimes confused with *potency*. Confusion can be avoided by reserving toxicity to refer to a chemical's intrinsic harmful properties (Chapter 1) and potency to refer to the median dose.

Efficacy, like potency, is a term that is more meaningful in the context of pharmacological therapeutics than toxic chemical risk. Efficacy refers to the percentage of a patient population that responds to drug treatment, i.e., the fraction of patients taking a drug who manifest the desired therapeutic endpoint. In Figure 5.5, the efficacies of drugs A, C, and D are similar. The efficacy of drug B is substantially lower; drug B is said to be less efficacious than drugs A, C, and D. Because of the way efficacy and potency are defined, a drug can be less efficacious than other drugs even though it is more potent (Drug B in Figure 5.5).

In the context of toxic chemicals, efficacy refers to the maximum frequency of a specified toxic effect. For example, one chemical may produce a toxic effect in 100% of an exposed population, while a second chemical produces the same toxic effect, but only in 65% of the population. The second toxic chemical is then said to be less efficacious than the first. The concept of efficacy, like potency, is more awkward when applied to toxic chemicals than therapeutic drugs. The persistence of these terms is a reminder that toxicology grew out of the field of medical pharmacology fairly recently, less than a century ago.

5.6 TOXICITY INVESTIGATIONS IN INDIVIDUAL ORGANISMS VS. POPULATIONS

By definition, toxicity testing is performed on populations of test organisms, not on individuals. The reason is that toxicity testing aims to determine the frequency of a specified toxic effect as a function of exposure in order to assess risk (Chapters 8 and 9). As stated previously, a toxicity test is, in effect, an epidemiological study of chemical disease under controlled laboratory conditions. The results are used to manage toxic chemical risk to human populations. Toxicity testing in nonmammalian species supports ecological risk assessments with the goal of protecting wildlife populations.

While full-scale toxicity testing is conceived as a basis for quantitative risk assessment and the protection of threatened populations, small-scale toxicity investigations in individual organisms or small groups of organisms can provide information that is valuable in assessing risk. Small-scale studies may be used to plan full-scale toxicity tests by identifying toxic effects and the dose ranges over which they are manifested. Subchronic toxicity testing is one example. Even smaller numbers of organisms may be used to investigate mechanisms of toxicity at the cellular and molecular level. Qualitative information on the types of toxic effects a chemical can cause at different doses and the mechanism(s) by which the chemical may act is of considerable value in characterizing risk (Chapters 8 and 9). Dose–effect curves can be constructed for individual organisms and are referred to as graded dose–effect curves to distinguish them from cumulative dose–effect curves, which are constructed for populations of test organisms (Figure 5.5). The difference is that in a graded dose–effect curve, the y-axis shows the increasing intensity of a chemical's effect on an individual organism with dose, whereas a cumulative dose–effect curve displays the increasing percentage of a population showing a specified toxic effect as a function of dose. Population studies and cumulative dose–effect curves are the vehicles of choice for assessing and managing toxic chemical risk.

5.7 USING ANIMALS TO SCREEN PERSONAL-CARE PRODUCTS: LOCAL IRRITATION AND SENSITIZATION TESTS

Several animal tests are designed to screen chemicals in medical devices as well as personal-care products for their ability to irritate the skin and eyes or to cause an allergic reaction. Irritation is due to local effects at the site of contact, effects that represent another exception to the general definition of toxicity as requiring uptake of a chemical into the bloodstream and distribution to distant organs and tissues. Sensitization is manifested locally but depends on the body's immune system. Test results are expressed as the presence or absence of the specified localized effect in response to a standard test dose of the chemical. Dosage groups and cumulative dose–effect analyses are not employed, with some exceptions. Irritation and sensitization tests are designed to screen the thousands of chemical products that are intended for contact with body surfaces. The ethics of irritation and sensitization tests are controversial. A number of groups consider them cruel and advocate that they be banned.

5.7.1 Skin Irritation Test

This test is usually performed on albino rabbits. A patch of fur is shaved off, and a standard amount of chemical (usually 0.5 gram of a solid chemical or 0.5 ml of a liquid chemical) is applied to each of four 1-in.^{2} sites on the skin. Each site is then covered with a patch to ensure contact of the chemical with the skin. The degree of dermal exposure at each skin site is varied by abrading (roughening) the skin and by using patches made of either gauze or plastic. After four hours, the patches are removed, and irritation is scored on the basis of redness, swelling, and corrosion of skin tissue.

5.7.2 Eye Irritation Test

This test, called the Draize test, is also performed in albino rabbits. A standard amount of test chemical is instilled in one eye, and the eye is scored for irritation at several time intervals. The second eye serves as a control. In recent years, a low-volume eye test (LVET) has been developed that uses one-tenth the volume of liquid in the traditional Draize test, reducing animals' suffering and yielding more consistent results.

5.7.3 Skin Sensitization Test

This test examines the potential of a chemical to cause contact dermatitis, which is a type of immune response known as a delayed hypersensitivity reaction. Traditionally, guinea pigs have been the test species of choice because the sensitivity of their skin is similar to human skin. The general procedure is to shave off a patch of fur and apply the test chemical to the skin a number of times over a period of two to four weeks. The rationale is that repeated applications allow the body's immune system to register the presence of the chemical. The chemical is then withheld for two to three weeks to give the immune system an opportunity to mount a response, after which the test chemical is again applied to the skin. A hypersensitivity reaction is indicated by the skin becoming red (erythematous) at the site of application.

In recent years, the mouse local lymph node assay (LLNA) has been approved as an alternative to the guinea pig test. It is based on the proliferation of lymphocytes (a type of immune cell) in lymph nodes draining the site of contact with the chemical. In the LLNA, the substance is applied to the mouse's ear on days 1–3 of the test but withheld on days 4 and 5 to give the immune system a chance to respond. On day 6, the mouse is injected in its tail vein with a small amount of a radioactive DNA base such as ^{3}H-thymidine (tritiated thymidine) to label newly formed immune cells. The mouse is sacrificed, its auricular lymph nodes (located near the ears) are cut out, and their radioactivity is measured. Increased radioactivity in the lymph nodes of treated animals compared with controls indicates that the test chemical sensitizes the immune system and can cause contact dermatitis. The LLNA is more objective than the traditional guinea pig test because the amount of immune cell proliferation can be determined as a function of dose.

5.8 REDUCING THE USE OF ANIMALS IN TOXICITY TESTING

There is growing interest in methods to evaluate chemical toxicity using fewer vertebrate animals or eliminating the use of animals altogether, when possible. Several intergovernmental agencies foster the development of alternatives to traditional test methods, such as the Interagency Coordinating Committee for the Validation of Alternative Methods (ICCVAM) in the United States and the European Centre for the Validation of Alternative Methods (ECVAM). Extensive information on traditional and alternative toxicity tests is available on the Web site at http://www.AltTox.org. A handful of nonanimal tests can help characterize toxicity; however, they are not able to replace animal toxicity tests.

5.8.1 Toxicity Testing in Single Cells

Prokaryotic (bacterial) cells and cultured eukaryotic (mammalian) cells have been enlisted in the effort to assess toxic chemical risk. Success has been limited, and it is not difficult to understand why: A cell is a drastically simplified surrogate of a multicellular organism such as a human being. The body is endowed with a broad array of organs and tissues, and it seems safe to assume that many things can happen in a body to decrease or increase a chemical's toxicity that may or may not happen inside a single cell. The limitations of single cells in mimicking the physiology of whole organisms make it impossible to rely on them as a basis for protecting public health. Nevertheless, studies on prokaryotic and eukaryotic cells provide information that can be useful in screening chemicals early in their development and avoiding the expense and ethical dilemmas of toxicity testing in animals.

5.8.2 Use of Bacteria to Screen Chemicals for Their Potential to Cause Cancer (Carcinogenicity)

The most important application of single-celled bacteria to chemical risk assessment is the Ames test. The Ames test is based on the understanding that one of the molecular mechanisms by which chemicals produce toxic effects is by altering the structure of the genetic material, DNA, at the level of the base pairs comprising the genetic code. Referred to collectively as mutations, these changes in individual letters of the genetic code can, if not repaired by the cell, lead to cancer. A mutation is believed to be capable of initiating a multistep process of carcinogenesis and to be a necessary, although not a sufficient, condition for cancer to develop (Chapter 7).

The Ames test uses a particular species of bacteria, *Salmonella typhimurium*, to screen chemicals for mutagenicity. The test is designed as follows: Mutant strains of the salmonella bacterium that have lost the ability to synthesize the essential amino acid histidine, designated *his*$^-$, have been isolated. These *his*$^-$ bacteria cannot grow and reproduce unless their food source contains histidine. If, upon exposure to the test chemical, descendants of these mutant strains again become able to grow on histidine-free medium, they will have undergone a mutation that cancels out the first mutation. The ability of a chemical to induce bacteria to undergo a so-called back mutation from a *his*$^-$ phenotype (unable to grow unless histidine is in the growth medium) to a *his*$^+$ phenotype (able to grow even if there is no histidine in the growth medium) is a direct measure of the chemical's mutagenic potential, i.e., its mutagenicity.

Some chemicals do not start out being mutagenic, but due to a process called toxication, they become mutagenic after they enter an animal's body (Chapter 6). As a result of a toxication reaction performed by a powerful enzyme in the liver or some other organ, a chemical's molecular structure may be transformed from innocuous to mutagenic. To assess a chemical's potential to undergo toxication to a mutagen, a modified Ames test is performed. An extract is prepared from rat livers, and the test chemical is incubated with the extract to give the liver enzymes a chance to alter its structure. The salmonella bacteria carrying the *his*$^-$ gene are then exposed to the test chemical after the chemical has been treated with rat liver extract, and the rate of back mutation to the *his*$^+$ phenotype is scored.

Mutagenicity in the Ames test indicates a chemical's general potential for DNA-related toxicity. In order to elucidate specific toxic effects, however, including carcinogenesis, a chemical that tests positive in the Ames test must also be tested in animals. A mutation in DNA is thought to be the main initiating event in the multistep process of carcinogenesis, and therefore the correlation between a chemical's mutagenicity in the Ames test and its carcinogenicity in animal tests is of great interest. Theoretically, this correlation should be 100%. The actual correlation is about 85%, excellent but not perfect. The remaining 15% of chemicals includes both false positives and false negatives. For example, manganese is mutagenic in the Ames test but is not carcinogenic in animal tests (false positive), while the estrogen analog diethylstilbestrol is not mutagenic in the Ames test but is carcinogenic in animal tests (false negative). The reasons for the less-than-perfect correlation between mutagenicity and carcinogenicity are not understood. The practical consequence is that the Ames test cannot be used to establish whether a chemical is or is not carcinogenic. Nevertheless, a positive result in the Ames test is very useful because it indicates an 85% probability that a chemical is carcinogenic. This information can be an important factor in deciding whether to abandon a new chemical or proceed with trying to bring it to market, including expensive toxicity testing in animals.

5.8.3 Use of Cultured Mammalian Cells to Screen for Genetic Toxicity

A wide variety of cells from animal and human tissues are routinely maintained in laboratory cultures and used for research and testing purposes. One use of mammalian cell lines is to screen chemicals for mutagenicity by applying genetic techniques similar to those used in the Ames test. Another, perhaps more important use of mammalian cells is to screen for forms of genetic toxicity that can affect higher organisms but not bacteria. In bacterial cells, which are referred to as prokaryotes, DNA exists more or less as a "naked" molecule. However, in the cells of higher organisms, referred to as eukaryotes, DNA is "packaged" in chromosomes inside a nucleus. When a eukaryotic cell divides (a process called mitosis), each of the two daughter cells receives the same complement of chromosomes and DNA, making them genetic replicas (clones) of their parent.

The situation is somewhat different in germ cells (eggs and sperm). The germ cells of eukaryotes divide in a process called meiosis, which is more complicated than mitosis because it includes additional steps that allow the DNA to undergo genetic recombination, with the potential for forging new combinations of genes inside the chromosomes. Egg and sperm cells end up with the same number of chromosomes and the same amount of DNA as their parent cells, but the combination of genes may be different. The shuffling of genes in meiosis increases a species' genetic diversity and hence its evolutionary potential.

Chromosome damage is a common form of genetic toxicity in higher organisms. Two examples of mammalian cell lines that are used to investigate chemical toxicity to chromosomes are Chinese hamster ovary cells and human lymphocyte cells. A chemical is investigated by adding it to the cell culture at a specific stage of the cell cycle. As the cells progress through the rest of their cycle and enter mitosis, their chromosomes condense (become visible under the microscope). The appearance of the chromosomes is compared with the appearance of normal chromosomes and evaluated for damage. Types of chromosomal damage include the loss of one or more pieces of a chromosome, the translocation of a piece to another chromosome where it does not belong, and extra or missing chromosomes. Chromosomes are much larger than DNA, and therefore chromosomal damage occurs on a larger physical scale than mutations in letters of the genetic code. Despite their difference in physical scale, however, point mutations in the DNA molecule and chromosomal aberrations that are visible under the microscope both serve as indicators of genetic risk. As such, they are useful tools in deciding whether to abandon development of an otherwise promising chemical or to proceed to toxicity testing in animals.

5.8.4 Structure/Activity Relationships

A few molecular features are considered to be strong indicators that a chemical is toxic. For example, chemicals with molecular structures that include an aromatic amine group, an amino azo dye structure, or a phenanthrene nucleus are considered to be potential carcinogens (Chapter 7). Chemicals with structural features related to valproic acid or retinoic acid are viewed as potential developmental toxicants. Referred to as structure/activity relationships (SARs), the presence in new chemicals of suspect molecular features is treated as a red flag, signaling potential risk and the need to either abandon further development of the new chemical or to subject it to toxicity testing. Unfortunately, only a handful of molecular structures are known to be reliable indicators of toxic chemical risk. Efforts to extend SARs as risk indicators using computerized methods, including a large study by the National Toxicology Program, have so far given disappointing results.

STUDY QUESTIONS

1. Describe the design of an acute-toxicity test. Define the LD_{50} derived from an acute-toxicity test.
2. Describe the design of a chronic toxicity test. Define the TD_{50} derived from a chronic toxicity test.
3. Which tends to be more conserved through evolution: (a) the doses of chemicals that cause toxic effects, or (b) the relative potencies of toxic chemicals? Explain how these two types of information might be combined to estimate the TD_{50} of a chemical, X, in humans from its TD_{50} in rats.
4. Describe three exposure scenarios in single-generation tests for reproductive toxicity. Name one toxic endpoint for each of the three exposure scenarios.
5. Define the maximum tolerated dose (MTD). How is the MTD used in the design of carcinogenicity tests in rats?
6. Discuss two types of toxicity studies that are used to determine the maximum acceptable toxicant concentration (MATC) in nonmammalian species such as fish and water fleas. How is the MATC defined?
7. Discuss the concept of threshold in chemical toxicity. How is the "no observed adverse effect level" (NOAEL) determined? What is the relationship of the NOAEL to the threshold of toxicity?
8. Explain how probit units may be converted to the percentages of a test population manifesting a specified toxic endpoint. Using information provided in the text as well as information that can be extracted from Figures 5.1 and 5.4, sketch a graph with probit units on the x-axis and percent on the y-axis. What percent of a population corresponds to 4.5 probits? To 5.8 probits?
9. What is the public health significance of the slope of a dose–effect curve? Describe two factors that could affect the slope.

10. Describe the Ames test for mutagenicity. Does the Ames test predict carcinogenicity? Chromosomal damage? Explain.
11. Define the following concepts as applied to therapeutic drugs and toxic chemicals: (a) potency, (b) efficacy, (c) toxicity.
12. What is a structure/activity relationship (SAR)? How useful have SARs proven to be in screening new chemicals for toxic chemical risk?

ANSWERS TO STUDY QUESTIONS

1. See Section 5.3.1 for acute-toxicity study design. The LD_{50} is the median lethal dose, which is defined as the dose of a test chemical, administered a single time at the start of the study, that results in the deaths of half of the test animals within the time frame of the study, usually 14 days. The route of exposure and the test species are included when reporting the LD_{50} because they have a significant effect on the lethal dose.
2. See Section 5.3.3 for chronic toxicity study design. The TD_{50} is the median toxic dose, which is defined as the dose of a test chemical, administered repeatedly, usually daily, that results in the manifestation of the specified toxic endpoint in half of the test animals within the time frame of the study, e.g., the adult life span of the test animal. As is the case with acute toxicity, the route of exposure and the test species also have a significant effect on the dose of a chemical causing chronic toxicity, and they are reported with the TD_{50}.
3. The relative potencies of toxic chemicals tend to be more conserved in evolution than the actual doses that cause toxicity. To estimate the TD_{50} of chemical X in humans, its TD_{50} can first be determined in rats and compared to the TD_{50} values of other chemicals that have been previously determined in rats and that also happen to be known in humans. As a numerical example, assume that the TD_{50} values for chemicals X and B in rats are 2×10^{-1} mg/kg and 6×10^{-4} mg/kg, respectively; in other words, X is 333 times less potent than B in rats. Now assume the TD_{50} for B in humans is estimated from data collected in the wake of an industrial accident to be 3×10^{-2} mg/kg. Since X is less potent than B in rats, it is likely, though not certain, that X is also less potent than B in humans. It would be overly simplistic to assume that X is 333 times less potent than B in humans, the same as in rats. However, the relative toxicities of X and B in rats can be used to ballpark the TD_{50} for X in humans at somewhere in the neighborhood of $(3 \times 10^{-2}) \times (333) = 10$ mg/kg.
4. See Section 5.3.4 for design of single-generation reproductive toxicity tests and for endpoints used in Segments I, II, and III.
5. The maximum tolerated dose (MTD) is chosen to result in the deaths of not more than half of the test animals in a chronic toxicity test. Statistical power is decreased when more than 50% of the test population dies. In a carcinogenicity test, the MTD is chosen such that deaths result only from cancer and not from other effects of high-dose exposure.

6. The MATC may be determined using a full life-cycle test, which entails exposure during the entire life cycle of the organism, or a partial life-cycle test, in which the organism is exposed during one stage of its life cycle because it is considered to be particularly vulnerable to chemical assault. The MATC is defined as the concentration of a chemical below which a population is able to maintain its size and above which a population declines and eventually disappears.
7. The threshold is the dose of a chemical below which a toxic effect does not occur and above which the toxic effect occurs with increasing frequency as the dose is increased. The threshold dose cannot be measured directly because it does not itself cause the toxic effect. The NOAEL is the dose that causes the toxic effect in enough animals to be observed but in too few to achieve statistical significance. The NOAEL is not a true threshold of toxicity; rather, it lies above the threshold. Approximately 5% to 10% of animals are affected by the NOAEL, presumably the most sensitive individuals in the test population.
8. Complicated mathematical equations are not required to approximate the relationship between probit units and percent. Draw a graph with percent on the x-axis and probits on the y-axis. Locate seven points corresponding to zero standard deviations and to ±1, ±2, and ±3 standard deviations. Connect the points to obtain an S-shaped graph. Read the percents corresponding to probits directly from the graph.
9. The steeper the slope in a dose–effect curve, the more members of an at-risk population are affected by each increase in exposure. A steep dose–effect curve indicates that a chemical has the potential to become a public health problem more quickly than a chemical with a shallow dose–effect curve. The slope is mainly a function of the toxic chemical itself. In addition, the strain of animals used in the toxicity test can affect the slope.
10. See Section 5.8.2 for a description of the Ames test. Approximately 85% of carcinogenic chemicals are detected by the Ames test, i.e., the correlation between the Ames test for mutagenicity, on the one hand, and carcinogenicity tests in animals, on the other, is about 85%. The Ames test produces both false positives and false negatives. It is not able to predict chromosomal damage because it is performed in prokaryotes, which lack chromosomes.
11. *Potency* is equivalent to the median dose (LD_{50}, TD_{50}, ED_{50}) in the context of a population study and a cumulative dose–effect curve. In the context of a graded dose–effect curve for an individual organism or an isolated tissue or cell extract, it is the dose that elicits 50% of the maximum physiological effect and is sometimes referred to as the 50% effective concentration, or EC_{50}. The lower the median dose, or the lower the 50% effective concentration, the more potent the chemical is said to be. Potency can refer to any physiological effect, including the therapeutic effect of a drug or the harmful effect of a toxic chemical. *Efficacy*, in the context of a population study, refers to the maximum percent of a population exhibiting a beneficial or a toxic effect. In the context of an individual organism or isolated tissue or extract, efficacy refers to the maximum physiological effect under

investigation. The term *toxicity* describes a specific class of harmful chemical properties (Chapter 1). Toxicity is not a synonym for potency.
12. SAR studies use structural features of known toxic chemicals to predict toxicity in new chemicals. This approach has succeeded for a small number of molecular structures. Unfortunately, ongoing efforts to identify additional red-flag structures have so far proven disappointing.

REFERENCES

AltTox.org. Non-animal methods for toxicity testing. http://www.alttox.org/
ECVAM. European Commission Joint Research Centre. Institute for Health and Consumer Protection, European Centre for the Validation of Alternative Methods (ECVAM). http://ecvam.jrc.ec.europa.eu/
ICCVAM. U.S. Department of Health and Human Services, National Toxicology Program, Interagency Coordinating Committee on the Validation of Alternative Methods (ICCVAM). http://iccvam.niehs.nih.gov/
Kamrin, M. A. 1988. *Toxicology: A primer on toxicology principles and applications*. Chelsea, MI: Lewis Publishers.
Katzung, B. G. 1992. *Basic and clinical pharmacology*. 5th ed. Norwalk, CT: Appleton & Lange.
Klaassen, C. D., ed. 2001. *Casarett and Doull's toxicology: The basic science of poison*. 6th ed. New York: McGraw-Hill.
Pepper, I. L., C. P. Gerba, and M. L. Brusseau, eds. 1996. *Pollution science*. San Diego: Academic Press.

SUGGESTED READING

Hodgson, E. 1987. Measurement of toxicity. In Hodgson, E. and P. E. Levy, *A textbook of modern toxicology,* pp 233–285. New York: Elsevier.

6 The Body's Defenses against Chemical Toxicity

6.1 INTRODUCTION

Chapters 2–5 introduced the science involved in tracking toxic chemicals and evaluating whether they are capable of producing adverse effects at the concentrations humans and other species are likely to encounter in their environments. Chemical partitioning, advective transport, and bioaccumulation are some of the conceptual tools for charting pathways of exposure, and the mass-balance concept offers a framework for organizing information about exposure pathways into a working model of environmental fate and transport (Chapter 2). The key to understanding toxic chemical risk is the dose–effect phenomenon, a form of the law of mass action, which makes it possible to characterize and quantify the relationship between chemical dose and toxic effect under controlled laboratory conditions (Chapter 3). The ability of toxic chemicals to harm human health and the environment can be investigated prospectively through toxicity testing in laboratory animals (Chapter 5). Human health impacts can be investigated retrospectively through epidemiological studies of chemicals possessing toxicities that may have eluded the regulatory process (Chapter 4).

While Chapters 2–5 introduced the science of chemical exposure and toxic effect, they considered biological organisms essentially as "black boxes." They did not describe the physiological processes that operate between the time an organism makes contact with a toxic chemical and the time—hours, days, weeks, or years after contact—when a toxic effect emerges. For most risk assessment and management purposes, the black-box approach to biological receptors works well, because it encompasses the most salient point: Toxic effect is related to chemical dose, and the lower the exposure, the less chance there is that a member of the exposed population will experience the toxic effect of concern. Just as driving a car does not require a detailed knowledge of the internal combustion engine, assessing and managing risk does not depend on a detailed knowledge of the biology of at-risk organisms.

Why, then, learn about the biology of toxicity? There are several reasons. The pathways chemicals travel in the body highlight toxic chemical risk and help bring it into focus. Chemicals' internal pathways reveal the body's defenses and the reasons some chemicals are detoxified while others are not. Appreciating the kinds of damage toxic chemicals can cause to cells helps in understanding the outer signs and symptoms of toxicity.

The goal of this chapter and the following chapter is to convey a working knowledge of the biology of toxicity. Extensive background in the life sciences, while helpful, is not essential. The emphasis is on concepts, and specialized terminology is kept to a minimum. The focus is on general physiological processes and molecular interactions that are broadly applicable to understanding how the body deals with toxic chemical risk.

6.2 EXPOSURE AND BIOAVAILABILITY

Before a toxic chemical can exert its effect, it first has to come in contact with an organism. Chemicals must then get into the bloodstream to produce a toxic effect, although there are exceptions to this general rule, such as locally acting lung and skin toxicants. For the vast majority of chemicals, however, penetration into the bloodstream is a prerequisite for toxicity. Chemicals that are unable to access the bloodstream are generally not able to produce toxic effects. Once in the bloodstream, a toxic chemical has to fend off the body's efforts to eliminate it until it has a chance to encounter vulnerable tissues and react with susceptible molecules inside cells.

It is helpful to use the terminology of risk assessment (Chapters 8 and 9) to describe exposure. Organisms that come in contact with toxic chemicals are generally referred to as "biological receptors" in risk assessment parlance. In the case of humans and other mammals, contact takes place in one of three ways: (a) by ingesting a chemical in food or water; (b) by inhaling a chemical in the air; or (c) by absorbing a chemical through the skin. These three forms of contact between a biological receptor and a toxic chemical—ingestion, inhalation, and dermal absorption—are referred to as the three major "routes of exposure." For aquatic species such as fish, uptake through the gills replaces inhalation as an exposure route.

Coming in contact with a toxic chemical is the first step of exposure. But contact alone is not enough. A second, crucial step is required: penetration of the toxic chemical from the digestive tract, lungs, or skin into the bloodstream. Once in the bloodstream, the chemical is quickly transported throughout the body. The fraction, or percent, of a toxic chemical that is able to enter the bloodstream is referred to as its bioavailability. The bioavailability of a chemical typically varies with the route of exposure and tends to be higher for inhalation and ingestion and lower for dermal contact.

Cell membranes have a major role in determining bioavailability. For example, when a chemical is ingested, it must diffuse through the membranes of cells lining the walls of the stomach and intestines to reach the capillaries and enter the wider circulation. The fraction that fails to make it out of the digestive tract is excreted in the feces. When the skin is exposed, a chemical must diffuse through several layers of epidermal cells before it encounters capillaries in the dermis—a particularly tall order for many chemicals. Access to the bloodstream is sometimes easiest for a chemical when it is inhaled, because the walls of the alveoli—the small, membranous sacks in the lungs where the exchange of O_2 and CO_2 takes place—are only one cell thick. For aquatic organisms, bioavailability via the gills depends on a chemical having a structure that enables it to diffuse across the membranes of gill cells into the gill capillaries. Capillaries are directly connected to veins in all species, and only seconds to minutes are required for a foreign chemical to reach the heart and be pumped into the general circulation. When toxic chemicals are ingested, they must pass through the liver on the way to the heart, and some are partially destroyed by protective liver enzymes en route (see Section 6.6).

The ease of passage across cell membranes depends on a chemical's size, electrical polarity, and lipophilicity (affinity for chemical environments characterized by fatty, carbon-rich molecules). The reasons for this selectivity lie in the structure and physical-chemical nature of cell membranes.

6.3 THE CELL MEMBRANE

The body is composed of trillions of cells working together in what might be thought of as a cellular symphony of everyday life. Cells are the smallest units of life, because they are capable of carrying out all of life's functions. Cells feed, metabolize, excrete wastes, reproduce, and interact with their environments. Indeed, many species of organisms consist of a single cell, for example, bacteria and yeast. Eukaryotic (higher) cells contain a variety of structures called organelles (Figure 6.1). Organelles, in turn, are composed of large biological molecules that carry out the vital functions of the cell. For example, the nucleus is an organelle that contains the genetic material, DNA, which codes for the assembly of proteins essential to the life of the cell and which is replicated by the cell when it divides. The mitochondria, which have DNA of their own and which are believed to be descended from primitive cells captured over a billion years ago by the ancestors of eukaryotic cells, transform chemical energy stored in food into ATP (adenosine triphosphate), the "energy currency" used by the cell. Some specialized cells such as spermatozoa possess flagella that enhance their motility. Cells depend on their organelles to carry out the diverse functions that are essential for the body to remain healthy. The nucleic acids, proteins, and other large molecules that comprise cells' organelles form the underlying fabric of tissues and organs that can be undermined by toxic chemicals.

The cell membrane, itself an organelle, plays a key role in guarding the body against toxic chemicals. The cell membrane is made up of a type of large molecule called phospholipids (Figure 6.2). Phospholipids have two distinct parts. The lipid part is composed of a carbon-rich fatty acid moiety, while the phosphate part contains a negatively charged phosphate moiety. The lipid part has a chemical affinity for other carbon-rich structures and is said to be lipophilic (Greek: "fat loving") or, alternatively, hydrophobic (Greek: "water-hating"). The electrically polarized phosphate part has a chemical affinity for other polar molecules such as water and is said to be hydrophilic (Greek: "water loving"). The hydrophobic "tails" of phospholipid molecules line up with each other and form the interior of cell membranes while their hydrophilic "heads" face outward to better interact with the watery environments inside and outside the cell. The resulting structure is called a lipid bilayer or "sandwich" in which the hydrophilic "heads" of phospholipid molecules are on the outside and their hydrophobic "tails" are on the inside of the membrane (Figure 6.2b).

The sandwich structure of cell membranes is stable because the association of phospholipid "tails" with each other and the association of phosphate "heads" with water are both energetically favorable. While the membrane is structurally stable, it is not a rigid shell. Large protein, carbohydrate, and lipid molecules are embedded in the cell membrane, and using special techniques, these structures can be observed

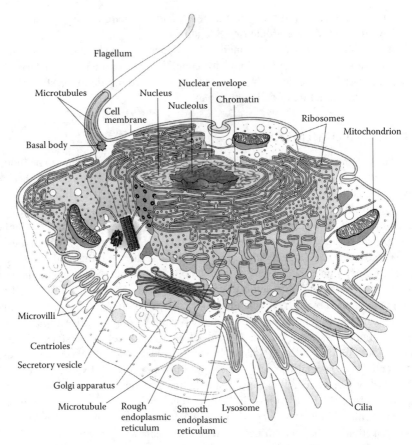

FIGURE 6.1 Composite diagram of a eukaryotic cell and its organelles. Cells contain a number of vital structures called organelles (not drawn to scale). Organelles are composed of large biological molecules such as proteins, lipids, carbohydrates, and nucleic acids (DNA and RNA). Not all cells have all the organelles shown in the diagram; for example, most cells lack cilia and flagella. (Reprinted with permission from David Shier, Jackie Butler and Ricki Lewis, *Hole's Human Anatomy and Physiology*, 7th ed. [Dubuque: Wm. C. Brown, 1996], 63.)

to move around while staying within the plane of the membrane (Figure 6.3). Their mobility indicates that the cell membrane possesses fluid qualities. Some proteins protrude on the inner and outer surfaces of the membrane and act as receptors for chemical messengers. For example, a receptor protein may bind a hormone on the outside of the cell and then, as the protein is activated in response to the hormone, trigger a hormone-mediated process inside the cell.

A note on terminology is in order: *Receptor* was originally used in biochemistry to describe large cellular molecules like proteins that interact with smaller molecules such as hormones. The term is now also used in risk assessments to refer to organisms or ecosystems that may be impacted by a toxic chemical or other stressor (see Section 6.2 and Chapters 8 and 9). Which of the two meanings of *receptor* is intended may be surmised from the context.

The Body's Defenses against Chemical Toxicity

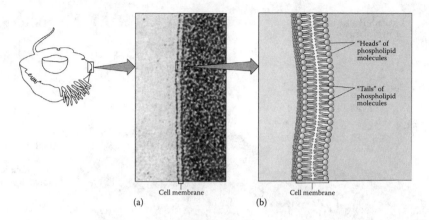

FIGURE 6.2 Structure of the cell membrane. (a) Transmission electron micrograph of a cell membrane at a magnification of 250,000. (b) Diagram of the cell membrane. Note the membrane's "sandwich" structure consisting of the phosphate-rich hydrophilic "heads" of phospholipid molecules on the outside and their hydrophobic lipid "tails" on the inside. The orientation of phospholipid molecules in the cell membrane is energetically favored, resulting in a stable structure. (Reprinted with permission from David Shier, Jackie Butler and Ricki Lewis, *Hole's Human Anatomy and Physiology*, 7th ed. [Dubuque: Wm. C. Brown, 1996], 66.)

Endogenous molecules, i.e., molecules that are normal constituents of the body, employ a number of mechanisms such as passive diffusion, facilitated diffusion, active transport, and pinocytosis to move back and forth across cell membranes. For the great majority of molecules that are foreign to the body, however, the only road to the bloodstream and the body's tissues and organs runs through the phospholipid bilayer of the cell membrane. Due to its unique structure, the bilayer is chemically selective. It imposes two general kinds of restrictions on the passage of foreign molecules: size and electrical polarity. Only molecules that are relatively small and that lack electrical polarity are able to squeeze past the outer layer of the "sandwich" and diffuse through its carbon-rich, hydrophobic interior. Molecules that are larger, that are electrically polarized, or both, are generally unable to make the passage. The cutoffs for size and polarity are not absolute. Molecules with a mass greater than approximately 600 daltons are generally but not always excluded. (Note: A single carbon atom has a mass of about 12 daltons.) Exclusion due to electrical polarity is also a matter of degree. The more a molecule contains polar atoms like oxygen and nitrogen, the more hydrophilic it is and the more it tends to be repelled by the hydrophobic chemical environment inside the membrane. Conversely, the more carbon atoms a chemical contains, the easier it is to get through the hydrophobic interior of the "sandwich."

One general indicator of a chemical's ability to cross cell membranes is its octanol-water partition coefficient, or K_{ow} (Chapter 2). The more lipophilic (hydrophobic) a foreign molecule is, i.e., the higher its K_{ow}, the more readily it crosses the cell membrane, enters the bloodstream, and contacts vulnerable cells. K_{ow} values are good predictors of the relative bioavailability of small foreign chemicals. If the foreign chemical is toxic, the K_{ow} is a fair predictor of internal exposure and therefore of the

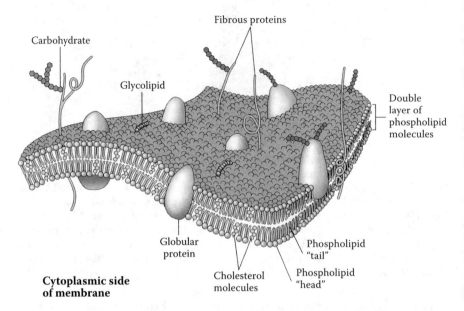

FIGURE 6.3 Fluidity of the cell membrane. Proteins embedded in the cell membrane are observed to move on the surface of the cell, providing evidence that the membrane's structure is not rigid but possesses some characteristics of a fluid. (Reprinted with permission from David Shier, Jackie Butler and Ricki Lewis, *Hole's Human Anatomy and Physiology*, 7th ed. [Dubuque: Wm. C. Brown, 1996], 66.)

probability that a toxic effect will occur. As a rule of thumb, the more lipophilic a toxic chemical, i.e., the greater its K_{ow}, the greater its bioavailability and the greater the likelihood that it may produce a toxic effect.

6.4 ELIMINATION BY THE KIDNEYS

The kidney is the principal organ for excreting foreign chemicals that have managed to book passage across cell membranes and enter the circulation. Kidney structure and function are outlined in Figure 6.4. Blood flows through the glomerular apparatus, where it is filtered under pressure through a semipermeable membrane barrier. The glomerular filtrate consists of the aqueous plasma and the relatively small molecules that are dissolved in it. Larger blood components such as proteins and red and white blood cells are excluded from the glomerular filtrate. The filtrate enters the renal tubule, which is in close contact with the small blood vessels in the kidney, the capillaries. Some foreign chemicals, or xenobiotics (from the Greek *xeno*, foreign, and *biotic*, living), tend to remain in the renal tubule and be excreted in the urine. Others tend to diffuse out of the renal tubule and into the interstitial fluid and the renal capillaries and reenter the general circulation. The extent to which foreign chemicals are confined to the renal tubule and excreted by the kidney is generally

The Body's Defenses against Chemical Toxicity

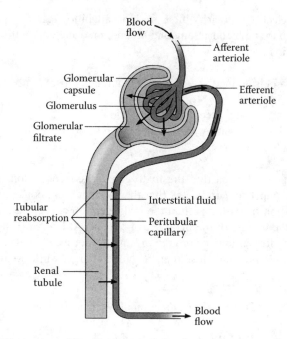

FIGURE 6.4 Elimination of foreign chemicals by the kidney. Chemicals that have been taken up into the bloodstream are filtered at the glomerulus and enter the renal tubule with the glomerular filtrate. Once in the renal tubule, some chemicals tend to stay in the tubule and be excreted in urine, while others tend to diffuse out of the renal tubule and be reabsorbed into the bloodstream. The electrical polarity of the foreign chemical plays an important role in determining whether the chemical is excreted or reabsorbed. (Reprinted with permission from David Shier, Jackie Butler and Ricki Lewis, *Hole's Human Anatomy and Physiology*, 7th ed. [Dubuque: Wm. C. Brown, 1996], 811.)

difficult to predict, with one notable exception: chemicals that happen to be weak acids or weak bases.

6.5 EXCRETION, ELIMINATION, AND WEAK ACIDS AND BASES

To set the stage for discussing the molecular characteristics that favor excretion of a foreign chemical by the kidneys, let us focus first on how exposure occurs following ingestion. In order to be taken up into the circulation, a foreign chemical must cross the walls of the stomach or intestines. The walls of these internal organs are made up of cells. Food molecules are broken down by digestive enzymes, and the breakdown products pass through the cells of the intestinal walls and into the bloodstream. In the case of toxic chemicals, uptake depends on whether a molecule's size and polarity allow it to pass through the membranes of the cells in the walls of the digestive tract. If a toxic chemical cannot leave the digestive tract, it remains outside the bloodstream, and true exposure does not occur. The chemical is excreted in the feces. Excretion by the kidney is not required.

A number of chemicals are either weak acids or weak bases. While this number is small compared with the total universe of chemicals, weak acids and bases illustrate the impact of electrical polarity, first, on bioavailability, e.g., on uptake from the digestive tract into the bloodstream, and second, on excretion by the kidney. Some definitions are in order:

1. A weak acid is defined as a molecule that readily loses a proton, or hydrogen ion (H^+).
2. A weak base readily accepts a hydrogen ion.
3. The pH is the negative logarithm to the base 10 of the hydrogen ion concentration.

Thus, a pH of 1 indicates that the hydrogen ion concentration is 10^{-1}, which is very acidic. A pH of 7 indicates that the hydrogen ion concentration is 10^{-7}, which is a million times lower than the hydrogen ion concentration at pH 1 and which lies in the middle of the pH scale of 1 to 14. At lower pH, the higher concentration of hydrogen ions favors the protonation of weak acids and bases (law of mass action). Weak acids lose their negative charge, while weak bases gain a positive charge:

$$A^- + H^+ \leftrightarrow AH \text{ (Protonation of weak acid at low pH)}$$

$$B + H^+ \leftrightarrow BH^+ \text{ (Protonation of weak base at low pH)}$$

where A is a weak acid, B is a weak base, and H^+ is a proton (hydrogen ion). At higher pH, the lower concentration of hydrogen ions favors the dissociation of protons from weak acids and bases (again, the law of mass action). As they lose their protons at higher pH, weak acids acquire a negative charge, while weak bases lose their positive charge and become electrically neutral:

$$AH \leftrightarrow A^- + H^+ \text{ (Deprotonation of weak acid at high pH)}$$

$$BH^+ \leftrightarrow B + H^+ \text{ (Deprotonation of weak base at high pH)}$$

The stomach is an acidic environment with a pH of about 1. The small intestine has a much more alkaline (basic) pH of around 6.4. The pH of the renal tubule is influenced by the constituents of the glomerular filtrate and ranges from slightly alkaline to slightly acidic. Depending on the dissociation constant of a weak acid or base and the pH of the tissue where it finds itself in the body, the concentration of uncharged molecules relative to charged molecules will change.

The ratio of charged to uncharged molecules as a function of pH may be calculated using the Henderson-Hasselbalch equation:

$$\log(\text{protonated}/\text{unprotonated}) = pKa - pH \tag{6.1}$$

where protonated refers to a weak acid or base containing a hydrogen ion, unprotonated refers to a weak acid or base from which the hydrogen ion has dissociated, pKa is the association (equilibrium) constant of the weak acid or base, and pH is

The Body's Defenses against Chemical Toxicity

the negative logarithm to the base 10 of the hydrogen ion concentration. If a foreign chemical is a weak acid or base and is ingested with food or water, it will be unable to penetrate into the bloodstream and will be excreted in the feces more or less to the extent that it carries an electrical charge preventing it from crossing the intestinal wall and entering the circulation. If it manages to enter the circulation, it will tend to be excreted by the kidney to the extent that the pH in the renal tubule causes it to have a negative or positive charge.

Aspirin and morphine exemplify the impact of pH on the uptake of weak acids and bases from the intestinal tract into the bloodstream and their excretion by the kidneys. Aspirin is a weak acid with a pKa of 3.5. In the acidic environment of the stomach with a pH of 1, solving the Henderson-Hasselbalch equation shows that there will be 316 times more uncharged (protonated) than negatively charged (unprotonated) molecules of aspirin. The uncharged aspirin is taken up from the stomach because it can cross the membranes of the cells lining the stomach wall and enter the bloodstream. Morphine, a weak base, is positively charged when it is protonated and electrically neutral when it is unprotonated. Solving the Henderson-Hasselbalch equation for morphine, which has a pKa of 7.9, shows that the ratio of charged (protonated) to uncharged (unprotonated) morphine molecules in the stomach is 7,943,282 to 1, dramatically lowering its bioavailability from the stomach. When morphine reaches the small intestine, where the pH is about 6.4, the ratio of charged to uncharged molecules drops to 32 to 1, still not particularly favorable for uptake, but enough for about 3% of the morphine molecules to enter the bloodstream.

If a person ended up with toxic levels of morphine in his or her bloodstream, it would be the kidney's job to eliminate the drug. A strategy that physicians sometimes employ to support the elimination of weak bases like morphine by the kidney is to acidify the urine, i.e., to make the glomerular filtrate slightly more acidic by administering a weak acid like ammonium chloride by mouth or intravenously in order to lower the pH in the renal tubule and promote protonation of the weak base and its excretion in urine. In the case of an aspirin overdose, supporting the patient might include alkalinizing the urine, i.e., administering a weak base like sodium bicarbonate to increase the percentage of negatively charged aspirin molecules that are trapped in the renal tubule and excreted in the urine.

6.6 BIOTRANSFORMATIONS

While the kidney is able to eliminate some xenobiotics on its own, many foreign chemicals are sufficiently small and nonpolar to pass through cell membranes and evade renal excretion. How does the body get rid of foreign chemicals that are taken up into the bloodstream and filtered at the glomerulus but then escape from the renal tubule and reenter the circulation?

The body possesses two great classes of enzymes that transform the structures of xenobiotic (foreign) molecules, making it more difficult for them to elude excretion. Most but not all of these enzymes are located in the liver. The first class of biotransformation enzymes, referred to collectively as catalyzing Phase I reactions, generally have the effect of adding oxygen atoms to susceptible chemicals (Table 6.1). The largest class of Phase I enzymes are called mixed function oxidases, or MFOs. The most

TABLE 6.1
Representative Phase I Biotransformation Reactions Catalyzed by the Cytochrome P-450 Family of Enzymes

Reaction Class	Structural Change	Drug Substrates
Cytochrome P-450–Dependent Oxidations:		
Aromatic hydroxylations	R-phenyl → R-phenyl-OH (via epoxide intermediate)	Acetanilid, propranolol, phenobarbital, phenytoin, phenylbutazone, amphetamine, warfarin, 17α-ethinyl estradiol, naphthalene, benzpyrene
Aliphatic hydroxylations	$RCH_2CH_3 \rightarrow RCH_2CH_2OH$ $RCH_2CH_3 \rightarrow RCHCH_3$ $\quad\quad\quad\quad\quad\quad\;\; |$ $\quad\quad\quad\quad\quad\quad\;\; OH$	Amobarbital, pentobarbital, secobarbital, chlorpropamide, ibuprofen, meprobamate, glutethimide, phenylbutazone, digitoxin
Epoxidation	$RCH=CHR \rightarrow R-\underset{H}{\overset{O}{C}}-\underset{H}{C}-R$	Aldrin
Oxidative dealkylation		
N-dealkylation	$RNHCH_3 \rightarrow RNH_2 + CH_2O$	Morphine, ethylmorphine, benzphetamine, aminopyrine, caffeine, theophylline
O-dealkylation	$ROCH_3 \rightarrow ROH + CH_2O$	Codeine, p-nitroanisole
S-dealkylation	$RSCH_3 \rightarrow RSH + CH_2O$	6-Methylthiopurine, methitural
N-oxidation		
Primary amines	$RNH_2 \rightarrow RNHOH$	Aniline, chlorphentermine
Secondary amines	$\underset{R_2}{\overset{R_1}{>}}NH \rightarrow \underset{R_2}{\overset{R_1}{>}}N-OH$	2-Acetylaminofluorene, acetaminophen
Tertiary amines	$\underset{R_3}{\overset{R_1}{\underset{|}{R_2-N}}} \rightarrow \underset{R_3}{\overset{R_1}{\underset{|}{R_2-N\rightarrow O}}}$	Nicotine, methaqualone
S-oxidation	$\underset{R_2}{\overset{R_1}{>}}S \rightarrow \underset{R_2}{\overset{R_1}{>}}S=O$	Thioridazine, cimetidine, chlorpromazine
Deamination	$RCHCH_3 \rightarrow R-\underset{NH_2}{\overset{OH}{C}}-CH_3 \rightarrow R-CCH_3 + NH_3$ $\quad\;\; |\quad\quad\quad\quad\quad\; |\quad\quad\quad\quad\quad\;\; \|$ $\quad NH_2\quad\quad\quad\; NH_2\quad\quad\quad\quad O$	Amphetamine, diazepam

TABLE 6.1 (CONTINUED)
Representative Phase I Biotransformation Reactions Catalyzed by the Cytochrome P-450 Family of Enzymes

Reaction Class	Structural Change	Drug Substrates
Desulfuration	$R_1\!\!>\!\!C\!=\!\!S \rightarrow R_1\!\!>\!\!C\!=\!\!O$ (with R_2)	Thiopental
	$R_1\!\!>\!\!P\!=\!\!S \rightarrow R_1\!\!>\!\!P\!=\!\!O$ (with R_2)	Parathion
Dechlorination	$CCl_4 \rightarrow [CCl\cdot_3] \rightarrow CHCl_3$	Carbon tetrachloride

Source: Modified from Bertram Katzung, *Basic and Clinical Pharmacology*, 5th ed. (Norwalk: Appleton & Lange, 1992), 53.

Note: Therapeutic drugs are listed as examples of foreign chemicals whose structures are modified by Phase I reactions, often by the addition of an oxygen atom. Some Phase I reactions, e.g., dechlorination, do not involve oxygen. A few Phase I reactions are catalyzed by enzymes outside the cytochrome P-450 family; one example is the conversion of ethyl alcohol to acetaldehyde (not shown).

---

important group of MFOs are the cytochrome P-450 family of enzymes (Table 6.1). Oxidation by Phase I enzymes usually does not confer an electrical charge; however, oxygen makes molecules more electrically polarized and therefore more hydrophilic and less able to cross cell membranes.

In addition to increasing polarity, oxidation also has the effect of priming molecules for further transformation by a second class of enzymes, which are referred to collectively as catalyzing Phase II reactions. Phase II reactions are also called transferase reactions, because they involve the transfer of specific moieties (groups of atoms) from endogenous donor molecules to foreign chemicals (Figure 6.5). The transferred moieties are both bulky and hydrophilic, resulting in a two-fold protective effect: The foreign chemical becomes physically less able to squeeze through cell membranes, and it tends to be chemically more repelled by the hydrophobic interior of the cell membrane. Phase II reactions are considered to be true detoxication reactions. Once encumbered with a bulky hydrophilic moiety, a chemical almost invariably loses its toxicity and is excreted by the kidney.

Most Phase I and Phase II reactions take place in liver cells. These cells, called hepatocytes, are organized in wedgelike lobules; the lobules are situated between blood vessels on one side and bile canals on the other (Figure 6.6). As blood moves past the lobules, small nonpolar xenobiotics that have been taken up into the bloodstream diffuse out of the blood vessels and into the hepatocytes. Most Phase I and Phase II enzymes are located in the endoplasmic reticulum, a highly folded membranous structure in the cytoplasm of the cell; a few enzymes are found in the aqueous cytoplasm itself and in mitochondria (Figure 6.1). Once a molecule has been enzymatically transformed, it is exported by the hepatocytes either to the blood vessels or to the bile canals bordering the lobules (Figure 6.6). Size is a factor: Smaller

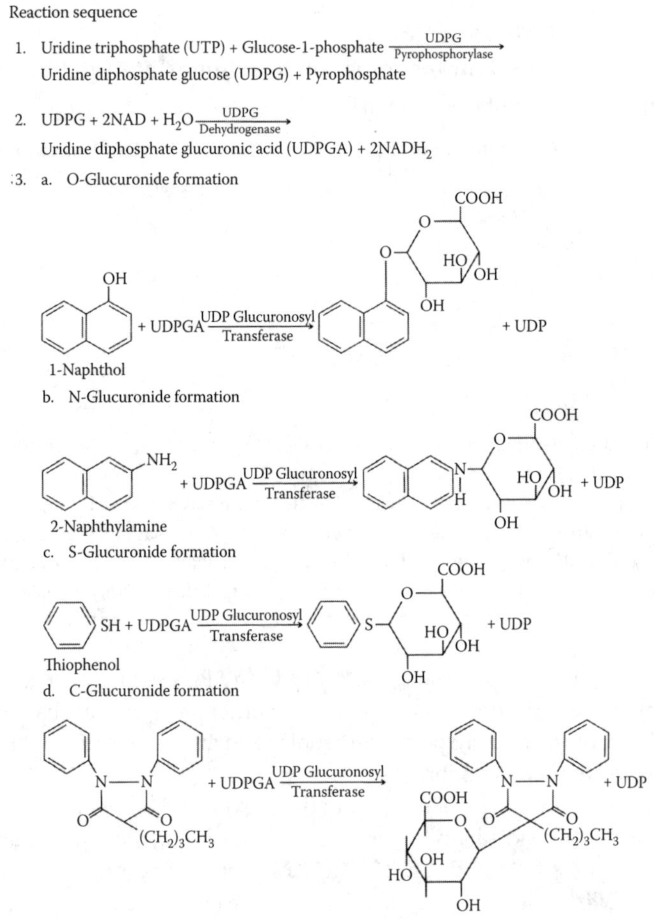

Reaction sequence

1. Uridine triphosphate (UTP) + Glucose-1-phosphate $\xrightarrow{\text{UDPG Pyrophosphorylase}}$
   Uridine diphosphate glucose (UDPG) + Pyrophosphate

2. UDPG + 2NAD + H$_2$O $\xrightarrow{\text{UDPG Dehydrogenase}}$
   Uridine diphosphate glucuronic acid (UDPGA) + 2NADH$_2$

3. a. O-Glucuronide formation

   1-Naphthol + UDPGA $\xrightarrow{\text{UDP Glucuronosyl Transferase}}$ [naphthyl O-glucuronide] + UDP

   b. N-Glucuronide formation

   2-Naphthylamine + UDPGA $\xrightarrow{\text{UDP Glucuronosyl Transferase}}$ [naphthyl N-glucuronide] + UDP

   c. S-Glucuronide formation

   Thiophenol (SH) + UDPGA $\xrightarrow{\text{UDP Glucuronosyl Transferase}}$ [phenyl S-glucuronide] + UDP

   d. C-Glucuronide formation

   [pyrazolidinedione with (CH$_2$)$_3$CH$_3$] + UDPGA $\xrightarrow{\text{UDP Glucuronosyl Transferase}}$ [C-glucuronide product] + UDP

FIGURE 6.5 Glucuronidation, an example of a Phase II biotransformation reaction. Uridine diphosphate glucuronic acid (UDPGA) is an endogenous molecule (i.e., it is native to the body) that is formed in two steps from uridine triphosphate, glucose-1-phosphate, and nicotinamide adenine dinucleotide. UDPGA is attached to a variety of small, relatively nonpolar foreign molecules such as 1-naphthol, 2-naphthylamine, and thiophenol by the Phase II enzyme UDP glucuranosyl transferase. Glucuronidation is one of many transferase reactions that take place in liver cells, encumbering foreign molecules with bulky hydrophilic moieties and causing them to be excreted in the urine and/or the feces. (Modified from E. Hodgson and P.E. Levi, *Textbook of Modern Toxicology* [New York: Elsevier, 1987], 74.)

molecules tend to be exported to the bloodstream, larger molecules tend to be exported to the bile canals, and intermediate size molecules tend to be exported to both blood and bile. Once in the bloodstream, biotransformed molecules are filtered at the glomerulus, trapped in the renal tubule, and excreted in urine. When biotransformed molecules are exported to bile, they move to the gall bladder, where bile is stored, and they are eventually excreted in the feces when bile is released from the gall bladder to the large intestine in response to a meal.

The Body's Defenses against Chemical Toxicity

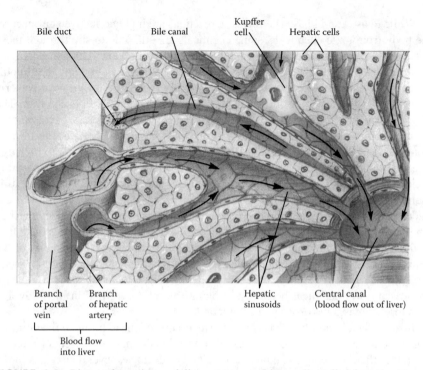

FIGURE 6.6 Biotransformation and liver structure. Liver cells called hepatocytes are wedged between blood channels (sinusoids) and bile canals. Foreign chemicals that have been taken up into the bloodstream diffuse down their concentration gradients out of the bloodstream and into hepatocytes. There they are subject to biotransformation by Phase I and Phase II enzymes. The structurally transformed chemicals move out of the hepatocyte, entering either the blood sinusoid (smaller chemicals) or the bile canal (larger chemicals) or both (chemicals of intermediate size). Biotransformed chemicals that enter the bloodstream are eliminated by the kidneys. Biotransformed chemicals that enter the bile canals are eliminated in the feces. (Reprinted with permission from David Shier, Jackie Butler and Ricki Lewis, *Hole's Human Anatomy and Physiology*, 7th ed. [Dubuque: Wm. C. Brown, 1996], 696.)

It should be noted that Phase I and Phase II enzymes protect the body from toxic chemicals in a manner that differs fundamentally from the immune system's protection of the body from pathogenic microorganisms. The immune system tailors antibodies to attack each invading microbe with "lock and key" specificity. By contrast, Phase I and II enzymes are relatively nonspecific. Each enzyme is capable of acting on a structurally diverse array of molecules, including endogenous chemicals that are part of the body's normal metabolic processes as well as foreign chemicals that pose a risk of toxicity. Each biotransformation enzyme effectively zeroes in on a specific functional group consisting of a small number of atoms while ignoring the larger three-dimensional structure of the molecule it is transforming. In doing so, it reduces the risk posed by a toxic chemical by blunting its toxicity and hastening its elimination from the body. The immune system does not act by promoting the excretion of invading microbes. Rather, it mobilizes microbe-specific antibodies to slow microbes' growth or kill them outright.

While Phase I and Phase II enzymatic reactions are highly effective at protecting the body from toxic chemicals, some chemicals are still able to slip through these defenses. One special case involves not the failure of Phase I and II enzymes to modify a foreign chemical, but the success of intestinal bacteria in undoing their modifications. The intestines of warm-blooded animals contain a vast array of bacteria living in symbiotic relationships with their animal hosts and with each other. The biotransformation, or metabolic, reactions performed by intestinal bacteria rival those of their animal hosts in scope and diversity. Occasionally, a biotransformation reaction performed by a Phase II enzyme is undone when bacteria in the large intestine happen to produce a class of enzymes that strip away the transforming moiety and render the foreign chemical small and nonpolar again. The chemical is then able to diffuse back out of the intestine and reenter the bloodstream, where it undergoes transformation again by Phase I and II enzymes in liver hepatocytes, then enters the bile canal, travels to the gall bladder and reaches the large intestine. Once in the intestine, it encounters the same bacteria, and they strip it once again of the detoxifying moiety it acquired in the hepatocytes. This process, called *enterohepatic circulation*, can be repeated many times. Enterohepatic circulation tends to increase the toxicity of a chemical because it prolongs its residence time in the body and therefore increases the exposure of vulnerable cells.

In addition to the failures of biotransformation reactions to promote the elimination of some xenobiotics, Phase I reactions occasionally have the effect of increasing rather than reducing a chemical's toxicity. For example, several enzymes in the cytochrome P-450 family are known to convert foreign chemicals into electrophiles. Electrophiles are molecules that are deficient in electrons; as such, they tend to react with molecules that are relatively electron-rich such as proteins, lipids, and DNA, which are essential to the health of cells. When a cellular molecule is attacked by an electrophile, its structure is altered, and its function is undermined (Chapter 7).

Two examples illustrate the toxication of foreign chemicals by enzymes in the cytochrome P-450 family. Tylenol (acetaminophen) is converted to a toxic electrophile by a Phase I reaction catalyzed by a cytochrome P-450 enzyme. The electrophile then goes on to react with proteins in liver cells. The reaction with liver proteins has the potential to destroy the liver and cause death in a matter of hours or days. The potentially lethal metabolite is generated every time a person takes Tylenol, but at the recommended dosage the concentration of the metabolite is kept low by a Phase II reaction, which conjugates the metabolite with the amino acid glutathione, a bulky detoxicating moiety. In the event of Tylenol overdose, however, the intracellular pool of glutathione becomes depleted, and liver cells cannot replenish it fast enough to prevent injury and death. A second example of toxication by a Phase I enzyme is aflatoxin, a natural product and a carcinogen. A cytochrome P-450 enzyme catalyzes the introduction of an expoxide moiety (Table 6.1), which renders the aflatoxin molecule electron-deficient. The resulting electrophilic aflatoxin metabolite goes on to react with DNA, with the potential to cause mutations and initiate a long-term (years to decades) process of carcinogenesis (Chapter 7).

While relatively rare compared with the large number of detoxication reactions performed by Phase I and II enzymes, toxication reactions pose significant risks of chemical disease. In recognition of their potential to increase the risk of cancer, toxication reactions have been incorporated into the Ames test for mutagenicity (Chapter 5). In the modified Ames test for toxication, a chemical is tested before and after exposure to an extract from rat liver hepatocytes. If the Ames test is negative for the chemical before incubation with the liver extract (i.e., the bacteria *Salmonella typhimurium* does not mutate from the his^- to the his^+ phenotype) but positive after incubation with the extract (the bacteria does mutate from his^- to his^+), then it may be concluded that the chemical has been rendered mutagenic by an enzyme in the liver.

6.7 THE KINETICS OF SINGLE-DOSE EXPOSURE: UPTAKE, DISTRIBUTION, AND ELIMINATION

The uptake of chemicals into the bloodstream, their distribution throughout the body, and their elimination by the liver and the kidneys have been described in qualitative terms. However, toxicologists like to put numbers on such processes: How quickly does the concentration of a toxic chemical build up? What level does it achieve? How long does it stay in the body?

Questions about the rapidity, extent, and duration of internal exposure to foreign chemicals are the province of the twin disciplines of pharmacokinetics and toxicokinetics. Pharmacokinetics is the study of foreign chemicals that act as therapeutic drugs. Their uptake, distribution, and elimination have been studied in detail in human subjects. Toxicokinetics is based on exactly the same principles as pharmacokinetics, but it is concerned with foreign chemicals with toxic, not therapeutic, effects. For obvious ethical reasons, toxicokinetic studies are limited to experimental animals. However, when dealing with the toxic ("side") effects of therapeutic drugs, the line between pharmacokinetics and toxicokinetics becomes blurred and effectively disappears.

How a foreign chemical moves into and out of the body may be divided into two broad categories: the kinetics of single-dose exposure and the kinetics of repeated-dose exposure. Examples of single-dose exposures are swallowing a tablespoon of Nyquil as a sleep aid or accidentally eating poisonous mushrooms on a walk in the woods. The kinetics of single-dose exposure may be investigated in clinical drug trials in people (therapeutic drugs) and in acute toxicity tests in laboratory animals (toxic chemicals). Examples of repeated-dose exposures are taking a prescription drug for high blood pressure and drinking water from an arsenic-contaminated well at home. Repeated-dose exposures may be investigated in clinical drug trials in people and in chronic and subchronic toxicity tests in laboratory animals.

Whether exposure is to a single dose or to repeated doses of a foreign chemical, kinetic studies focus on the concentration of the chemical in blood. As we have seen, blood has a pivotal role in the uptake, distribution, and elimination of foreign chemicals. Further, from a practical standpoint, blood concentrations of chemicals are relatively easy to measure—much easier, for example, than their concentrations in organs and cells. Because blood is relatively accessible, and because it plays a pivotal role in exposure, the waxing and waning of a chemical's concentration in blood define its kinetics.

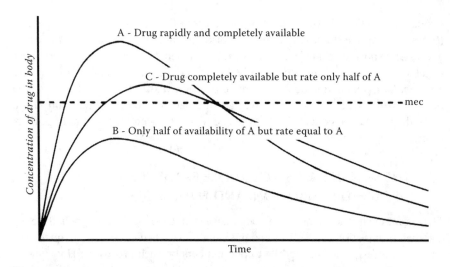

FIGURE 6.7 Blood concentration-time curves for three foreign chemicals. The length of time that blood concentrations exceed the minimum level of effectiveness (dashed line) depends on the balance among rate of uptake, bioavailability, and rate of elimination. The dashed line can be taken to represent the minimum blood concentration that is therapeutic (if the foreign chemical is a drug, as in this figure) or harmful (if the foreign chemical is toxic). See text for details. (Reprinted with permission from Bertram Katzung, *Basic and Clinical Pharmacology*, 5th ed. [Norwalk: Appleton & Lange, 1992], 41.)

Let us take a look at how blood concentrations change following a single dose of a drug. Not surprisingly, the uptake, distribution, and elimination of a foreign chemical overlap considerably in time. At each point in time, the blood concentration reflects a balance among these three distinct physiological processes. Blood concentration-time curves for three hypothetical drugs, A, B, and C, which have similar potencies and therapeutic effects and which are administered by mouth as a single dose, are shown in Figure 6.7. The drug must achieve a certain minimum concentration in blood in order to be therapeutically effective (dashed line in Figure 6.7). Reaching and maintaining that minimum concentration depends on the kinetics of each drug. Drug A is 100% bioavailable, and it is also taken up quickly. The result is a relatively long period of therapeutic effectiveness. Drug B is only 50% bioavailable, and elimination is sufficiently fast that it never reaches a therapeutic concentration in blood. Drug C, like A, is 100% bioavailable, but it is taken up at only 50% the rate of A. The slower uptake combined with ongoing elimination results in a reduction in the duration of therapeutic effectiveness by almost half. The same kinetic constraints—bioavailability, rate of uptake, and rate of elimination—also apply to blood levels of toxic chemicals. The only difference is that if A, B, and C were toxic chemicals instead of therapeutic drugs, the dashed line in Figure 6.7 would represent their minimum toxic concentration instead of their minimum therapeutic concentration. In other words, the balance among uptake, distribution, and elimination of a toxic chemical determines whether its concentration in blood is high enough to result in significant exposure and injury to cells.

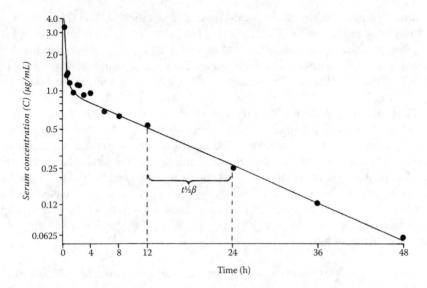

FIGURE 6.8 Blood-concentration time curve following intravenous administration of a foreign chemical. The antianxiety drug chlordiazepoxide (Librium) was administered as a single dose injected into a vein. The blood concentration decreased in two phases. The initial steep phase corresponds to the drug's rapid distribution within the body. The second, shallower phase corresponds to the elimination of the drug from the body by the combined actions of the liver and the kidneys. The slope of the shallow phase is equal to the elimination rate constant, k (Equation 6.4). The half-life may be determined from k (Equation 6.5). The half-life may also be determined directly from the graph as the time required for the blood concentration to decrease by a factor of two between any two points on the shallow portion of the curve. Note the logarithmic scale on the y-axis. (Modified from Bertram Katzung, *Basic and Clinical Pharmacology*, 5th ed. (Norwalk: Appleton & Lange, 1992], 8.)

When a foreign chemical gains access to the body by one of the three routes of exposure, i.e., ingestion, inhalation, or skin contact, its blood concentration-time curve reflects the combined effects of uptake, distribution, and elimination (Figure 6.7). A second approach to investigating the kinetics of a foreign chemical is to administer a single dose intravenously. The advantage of this approach is that it bypasses the process of uptake, and the chemical immediately becomes 100% bioavailable. The blood concentration-time curve now reflects the processes of distribution and elimination only (Figure 6.8). Distribution is reflected in the initial steep portion of the curve, while elimination is reflected in the second, shallower portion. Consider first the process of distribution.

Chemicals that are taken up into the bloodstream are distributed into various physiological compartments of the body. Examples of physiological compartments are water, fat, and bone. Physiological compartments do not refer to specific organs or anatomical structures. Rather, they are based on the physical-chemical characteristics of tissues, regardless of where in the body the tissues are located. Thus, the "fat compartment" refers to all fatty tissues everywhere in the body, the "water compartment" refers to the many parts of the body made up of water, and the "bone compartment" refers to all bones. A chemical's distribution into physiological compartments is strongly influenced by its physical-chemical properties. A chemical's aqueous solubility and its K_{ow} value (Chapter 2) are fair predictors of its general pattern of distribution. Hydrophilic chemicals that have high aqueous solubility and low K_{ow} values tend to be distributed in body water. Conversely, lipophilic (hydrophobic) chemicals that have low aqueous solubility and high K_{ow} values tend to be distributed in body fat.

The initial steep portion of the time-concentration curve in Figure 6.8 shows how quickly a chemical is distributed and reaches a steady state between its concentration in tissues and its concentration in blood. The second, shallower portion of the curve can be used to determine the volume of distribution, the total volume of all the physiological compartments in which a chemical is distributed. Just as physiological compartments of the body are virtual compartments, a chemical's volume of distribution, V_d, is a virtual volume. The V_d is defined as the ratio of the total amount of chemical in the body divided by the concentration of the chemical in the plasma (watery) fraction of blood:

$$V_d \text{ (liters)} = A_b \text{ (milligrams)}/C_p \text{ (milligrams/liter)} \qquad (6.2)$$

where V_d is the volume of distribution of the chemical, A_b is the amount of chemical in the body, and C_p is the concentration of the chemical in plasma. The V_d is determined by extrapolating the shallow portion of the curve in Figure 6.8 back to its point of intersection with the y-axis, which corresponds to the plasma concentration at time t = 0. This is the concentration it is assumed the drug would have had if it had been distributed instantaneously upon intravenous injection. The amount of drug in the body at t = 0 is the same as the dose of drug that was administered intravenously. Substituting the values for A_b and C_p at time t = 0 into Equation (6.2) gives V_d.

When interpreting the volume of distribution, it's important to keep in mind that it is a virtual volume, not a real one. V_d is defined as being inversely related to a chemical's concentration in blood. The more a chemical tends to leave the bloodstream, the larger its volume of distribution tends to be, by definition. Some chemicals are mostly or partly restricted to the bloodstream. They have volumes of distribution in the 7–15-liter range, consistent with the circulatory system comprising approximately 8% of body volume. (Note: Body volume in liters is roughly equal to body weight in kilograms. A kilogram is 2.2 pounds. A person weighing 155 pounds or 70 kilograms has a body volume of about 70 liters.) Some chemicals tend to be distributed in total body water, which includes the circulatory system as well as extracellular and intracellular water. Such chemicals typically have V_d values in the 40–100-liter range, consistent with approximately 60% of the human body consisting of water. More lipophilic (hydrophobic) chemicals with relatively large K_{ow} values tend to partition into body fat and typically have V_d values greater than 100 liters. Examples of these three general ranges of V_d values are aspirin, which has a volume of distribution of 11 liters; lithium, with a V_d of 55 liters; and chloroquine, the antimalarial drug, with a V_d of 13,000 liters. These volumes of distribution, like others in the medical literature, are based on a 70-kg male. The volume of distribution can vary considerably from person to person depending on body weight and relative amounts of lean and fatty tissue.

The Body's Defenses against Chemical Toxicity

How a chemical is distributed in the body affects the rapidity of its elimination. Chemicals that are distributed in body water are more immediately accessible to the bloodstream and hence are more available for elimination by the kidneys and the liver: As their blood concentration falls as a result of elimination, water-soluble chemicals diffuse out of body water and back into the bloodstream. On the other hand, chemicals that are distributed in fat or bone tend to be less immediately accessible to the bloodstream, and the time required for their elimination tends to be longer. Indeed, fat and bone have the potential to act as internal sources of low-level exposure over time as the concentration of a chemical builds up in these tissues. Examples are lead, which tends to be sequestered in bone, and persistent organic pollutants (POPs) like dioxin and polychlorinated biphenyls (PCBs), which tend to be sequestered in fatty tissue. These and other chemicals may persist in the body for months or years.

While the initial phase of the blood concentration-time curve of an intravenously administered chemical reflects its distribution into physiological compartments, the shallow phase reflects its elimination by the kidneys and/or the liver. Despite its physiological complexity, the elimination of many (though not all) chemicals from the bloodstream follows first-order kinetics. This means that equal fractions of a chemical disappear from the blood per unit time. A first-order kinetic process is described by an exponential equation:

$$[A_t/A_0] = e^{-kt} \tag{6.3}$$

where A_t is the amount of chemical in the body at time t following intravenous administration, A_0 is the amount of chemical in the body at time t = 0, k is the elimination rate constant, and t is the time elapsed since intravenous administration. Taking the natural logarithm of both sides of Equation (6.3):

$$\ln [A_t/A_0] = -kt \tag{6.4}$$

Equation (6.4) corresponds to the mathematical form of the second phase of the blood concentration-time curve in Figure 6.8, which is a semilogarithmic plot, i.e., it plots the logarithm of changing blood concentration against time. The slope of the second phase of the curve in Figure 6.8 is defined by the elimination rate constant, k. The larger the elimination rate constant k, the steeper the slope and the faster the chemical is eliminated from the bloodstream.

The rate of elimination can also be expressed in terms of a chemical's half-life. The half-life is inversely related to the rate of elimination: The larger the elimination rate constant k, the shorter is the half-life. The half-life can be calculated from the elimination rate constant by solving Equation (6.4) for $[A_t/A_0] = 1/2$:

$$\ln ([A_t/A_0]) = -kt$$
$$\ln 0.5 = -kt_{1/2}$$
$$0.693 = kt_{1/2}$$
$$t_{1/2} = 0.693/k \tag{6.5}$$

Equation (6.5) can be used to calculate the half-life of a chemical in the bloodstream when the first-order elimination rate constant is known. Conversely, it can be used to calculate the first-order elimination rate constant when the half-life is known. The half-life can also be determined graphically from the time-concentration curve in Figure 6.8 by picking any two serum concentrations in the curve's second phase that differ by a factor of two, then determining the difference between the times on the x-axis that correspond to the two concentrations. A chemical's half-life generally reflects the combined rate of its clearance from the bloodstream by the kidneys and the liver; for some chemicals, other organs may contribute to elimination, as well, for example, the lungs. The faster a chemical is cleared, the shorter is its half-life.

6.8 THE KINETICS OF REPEATED-DOSE EXPOSURE

In repeated-dose exposures, chemicals have a chance to accumulate in the body. The level a chemical reaches in the body depends on how large and frequent the doses are as well as on the chemical's disposition in the body, i.e., its bioavailability, volume of distribution, and rate of elimination. The blood level of a chemical at any point in time in effect reflects a steady state between the amount of chemical that enters the body and the amount that leaves. The principle of mass balance (Section 2.8) can be used to estimate body burden:

$$\text{Mass of chemical entering body} = \text{Mass of chemical exiting body} \quad (6.6)$$

The mass balance can be expressed using kinetic parameters:

$$(\text{Dose/Dosing interval}) \times \text{Bioavailability} = (\text{Amount in body})/\times k \quad (6.7)$$

The derivation of Equation (6.7) involves the concept of clearance and is beyond the scope of this book. Nonetheless, Equation (6.7) can be used to explain the factors that control the steady-state level of a chemical in the body. Thus, the left side of the equation states that the rate at which a chemical enters the body, meaning the bloodstream, is equal to the dosing rate (dose/dosing interval) multiplied by its bioavailability (fraction of a dose that is taken up into the bloodstream). Higher doses, shorter dosing intervals, and greater bioavailability result in more chemical being taken up per unit time. The right side of the equation states that the rate at which a chemical is eliminated is equal to the amount in the body multiplied by the elimination rate constant. In other words, consistent with the principle of mass action, the more chemical there is in the body, the greater the mass of chemical that exits the body per unit time. Rearranging Equation (6.7) to solve for the amount of chemical in the body at steady state and substituting $1.44 \times t_{1/2}$ for $1/k$ (Equation 6.5):

$$\text{Amount in body} = [(\text{Dose/Dosing interval}) \times \text{Bioavailability}] \times 1.44 \times t_{1/2} \quad (6.8)$$

Equation (6.8) shows that increasing the dose and decreasing the dosing interval result in increased accumulation of a foreign chemical in the body. This jibes with common sense. Equation (6.8) shows that a chemical's half-life also has a profound

The Body's Defenses against Chemical Toxicity

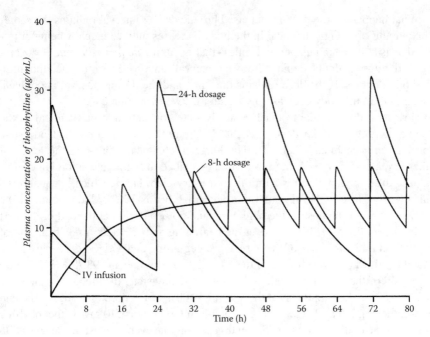

FIGURE 6.9 Idealized blood concentration-time curves of a foreign chemical resulting from a regular schedule of exposure. The drug theophylline is administered to an asthma patient by mouth every eight hours at a dose of 340 mg or every 24 hours at a dose of 1,020 mg. The average amount in the body (510 mg) and average plasma concentration (15 µg/ml) are the same under both dosing regimens (see Equation 6.8 and text). However, the large concentration swings at the higher dose pose the twin risks of adverse side effects (at the start of the dosing interval) and therapeutic ineffectiveness (at the end of the dosing interval). In a hospitalized patient, it may be advisable to avoid concentration swings by administering the drug intravenously (smooth curve). When exposure is to varying doses of a toxic chemical in the environment and it occurs on a random basis, blood levels can be expected to fluctuate much more than they do in a patient taking a prescription drug. (Reprinted with permission from Bertram Katzung, *Basic and Clinical Pharmacology*, 5th ed. [Norwak: Appleton & Lange, 1992], 44.)

effect on accumulation: The longer a chemical's half-life, the more it tends to build up in the body on repeated exposure.

The simplest case of repeated dose exposure is the administration of a therapeutic drug (Figure 6.9). When dosing occurs on a regular schedule, plasma concentration is characterized by a peak at the start and a trough at the end of the each dosing interval. If the dose is increased, the average amount in the body and the average concentration in plasma remain the same, provided the dosing interval is also increased (Equation 6.8), but peak plasma concentrations are higher and trough concentrations are lower. Large swings risk toxic side effects at high concentrations and therapeutic ineffectiveness at low concentrations. These twin risks are avoided when a drug is administered by continuous intravenous infusion such that the amount of drug in the body is maintained at a constant level (Figure 6.9).

What happens if the dose is increased but the dosing interval remains the same? The amount of foreign chemical in the body increases until as much is being eliminated as is being taken up per unit time—that is, until *rate in* = *rate out*. For example, if the dose is doubled and the dosing interval remains the same, the amount in the body doubles. If the dose remains the same and the dosing interval is doubled, the amount in the body decreases by a factor of two (Equation 6.8).

How long does it take to reach a steady-state concentration of chemical in the body assuming a regular dosing schedule? The time it takes to reach steady state depends on a chemical's half-life: The longer a chemical's half-life, the longer it takes to reach steady state. In the situation described in Figure 6.9, a foreign chemical (the drug theophylline) is administered every eight hours, which is equal to its half-life; the dose is 340 mg; and the bioavailability is 100%. The amount in the body is 100% of the dose at the beginning and 50% (170 mg) at the end of the first dosing interval; 150% of the dose at the beginning and 75% (255 mg) at the end of the second interval; 175% of the dose at the beginning and 87.5% (297.5 mg) at the end of the third interval; 187.5% of the dose at the beginning and 93.75% (318.75 mg) at the end of the fourth interval, and so on. At steady state, the amount in the body is 200% of the dose (680 mg) at the start and 100% of the dose (340 mg) at the end of each dosing interval. The average amount in the body is 510 mg. If theophylline has a volume of distribution of 35 L in this patient, the average plasma concentration is 15 µg/ml (Figure 6.9 and Equation 6.2). True steady state requires that *rate in* = *rate out*. For practical purposes, however, an average concentration equal to 90% of the plateau value is generally considered to be a reasonable approximation to steady state. It takes 3.3 half-lives for a foreign chemical to reach 90% of its average steady-state amount in the body, assuming exposure occurs on a regular basis.

Estimating body burdens of therapeutic drugs administered on a fixed schedule is one thing. Estimating body burdens resulting from random exposure to toxic chemicals is quite another. In addition to the random nature of exposure, accurate information on bioavailability and half-life are often not available. As a result of multiple uncertainties, estimates of human body burdens of toxic chemicals usually amount to educated guesses based on physical-chemical characteristics such as K_{ow} values, results of toxicity testing in animals, and accidental poisoning in humans.

A more direct approach to the question of body burden is to collect and analyze tissue specimens such as blood, urine, or hair from members of an at-risk population. Tissue analysis offers incontrovertible evidence of exposure and a snapshot of chemical levels. It is particularly valuable when the tissue is blood. However, tissue analysis does not provide a history of exposure over time, which is the foundation of risk assessment.

Basing risk assessment on toxicity testing in animals (Chapter 5) and providing default assumptions about exposure (Chapter 8) avoids the many pitfalls of toxicokinetics as a strategy for evaluating impacts of chemicals on human health. Information about a chemical's toxicokinetic properties—particularly its bioavailability and its half-life—can and do augment assessments of risk. However, toxicity testing in animals, not exposure and disease in humans, provides the foundation for managing risk from toxic chemicals.

The Body's Defenses against Chemical Toxicity

STUDY QUESTIONS

1. Define the following terms: (a) routes of exposure, (b) bioavailability, (c) elimination, (d) excretion.
2. Describe the structure of the cell membrane using the following terms: phospholipids, hydrophilic, hydrophobic, bilayer, "sandwich," polarity, selective, fluidity.
3. What is the hydrogen ion concentration when pH = 4.3? pH = 8.7?
4. A child is admitted to the hospital emergency room with a diagnosis of accidental aspirin overdose. Among other supportive measures, the physician considers administering a weak solution of sodium bicarbonate intravenously. What would be the rationale for administering sodium bicarbonate? If the pH of the patient's urine is 7.5, what percentage of the aspirin molecules are charged? Assuming administration of sodium bicarbonate raised the urine pH to 8.5, how much would the percentage of charged aspirin molecules increase or decrease?
5. What are the main characteristics of Phase I biotransformation reactions? What is the major class of enzymes that perform Phase I reactions? Where are Phase I enzymes located? What are the main characteristics of Phase II reactions? What is the major class of enzymes that perform Phase II reactions, and where are these enzymes located?
6. Describe in general terms the physiological processes of uptake, distribution, and elimination of a foreign chemical. Are these processes sequential? Do they overlap?
7. What is the volume of distribution of a foreign chemical? Estimate the volume of distribution of chlordiazepoxide from Figure 6.8.
8. The elimination rate constant is 3.9 day^{-1} for foreign chemical Y and 0.01 day^{-1} for foreign chemical W. What are these chemicals' half-lives in the body?
9a. What is the steady-state amount of chemical Y (from question 8) in the body if exposure occurs three times a week, the dose is 0.5 mg, and the bioavailability is 15%? How long will it take to reach steady state?
  b. What is the steady-state amount of chemical W (from question 8) in the body if exposure occurs once a week, the dose is 0.05 mg, and the bioavailability is 5%? How long will it take to reach steady state?

ANSWERS TO STUDY QUESTIONS

1. (a) There are three routes of exposure in humans: ingestion, inhalation, and dermal contact. Exposure routes are effectively the same in other land-dwelling creatures as in humans. Instead of inhalation, aquatic organisms are exposed through their gills. (b) Bioavailability refers to the percent of a dose of a foreign chemical that is taken up into the bloodstream. (c) Elimination refers to the set of physiological processes by which foreign chemicals that have penetrated into the bloodstream are removed from the body. (d) Excretion refers to the incorporation of a foreign chemical into

urine or feces. Excretion is the final step in the elimination of a foreign chemical from the bloodstream and the body. Fecal excretion can also refer to a chemical that passes through the intestinal tract without penetrating into the bloodstream.

2. The cell membrane is composed of large phospholipid molecules. Each molecule has a "head" region that contains phosphorus and oxygen atoms and is hydrophilic ("water-loving") and a "tail" region that contains mostly carbon atoms and is lipophilic ("fat-loving") or hydrophobic ("water-hating"). The phospholipids are oriented "tail"-to-"tail" in the cell membrane, an arrangement that is energetically favored in the aqueous (watery) environment of the body. The resulting membrane structure consists of two layers of phospholipid molecules and is referred to as a phospholipid bilayer or "sandwich." The cell membrane is stable but not rigid. It possesses some characteristics of a fluid, as indicated by the mobility of cell surface proteins. The cell membrane is selective with respect to the size and polarity of molecules that can diffuse across it to enter and leave the cell. In general, molecules larger than roughly 600 daltons are excluded. Smaller molecules may be partially or wholly excluded if they contain polar atoms such as oxygen, phosphorus, and nitrogen and/or if they carry a positive or negative electrical charge.

3. If pH = 4.3, then $[H^+] = 10^{-4.3} = 1/10^{4.3} = 1/19,953 = 5.01 \times 10^{-5}$.
   If pH = 8.7, then $[H^+] = 10^{-8.7} = 1/10^{8.7} = 1/501,187,234 = 2.0 \times 10^{-9}$.

4. Rationale for administering sodium bicarbonate: Aspirin is a weak acid, pKa = 3.5 (see text). Alkalinizing the urine by administering a weak base like sodium bicarbonate promotes the excretion of weak acids like aspirin. The Henderson-Hasselbalch equation can be used to calculate the fraction of aspirin molecules in charged and uncharged forms.

   At pH = 7.5:
   Log (Protonated/Unprotonated) = pKa − pH = 3.5 − 7.5 = −4.0
   Antilog (−4.0) = 10^{-4}; Protonated/Unprotonated = 10^{-4}

   There is 1 protonated, or uncharged, aspirin molecule for every 10,000 unprotonated, or charged, aspirin molecules. The percentage of charged aspirin molecules is therefore 99.99%.

   At pH = 8.5:
   Log (Protonated/Unprotonated) = pKa − pH = 3.5 − 8.5 = −5.0
   Antilog (−5.0) = 10^{-5}; Protonated/Unprotonated = 10^{-5}

   There is 1 protonated, or uncharged, aspirin molecule for every 100,000 unprotonated, or charged, aspirin molecules. The percentage of charged aspirin molecules is therefore 99.999%. Given the tiny percentage increase in charged aspirin molecules, alkalinizing the urine may not be an effective strategy for treating aspirin overdose.

5. Phase I biotransformation reactions typically add an oxygen atom to a foreign molecule. Oxygenation tends to have a dual effect: It makes the foreign molecule more polar so that it is somewhat less able to diffuse across cell membranes and therefore is somewhat more prone to elimination by the kidneys; and it primes the foreign molecule for a Phase II biotransformation

reaction. The biggest class of Phase I enzymes is the cytochrome P-450 family of enzymes, which are found in the endoplasmic reticulum of liver hepatocytes. Some Phase I enzymes are also found in the cell cytoplasm. Phase II biotransformation reactions typically add a bulky hydrophilic moiety to a foreign molecule. This has the dual effect of greatly reducing toxicity by altering a molecule's structure while promoting elimination by making it much more difficult for the molecule to escape excretion by the kidneys. Phase II reactions are considered true detoxication reactions. Most Phase II enzymes are classified as transferases because they transfer bulky hydrophilic moieties, e.g., glucuronic acid, from endogenous donor molecules to foreign molecules. Like Phase I enzymes, Phase II enzymes are located predominantly, although not exclusively, in liver hepatocytes.

6. Uptake is the process by which a foreign chemical enters the bloodstream. For the great majority of foreign chemicals, uptake depends on the ability to diffuse across cell membranes, which in turn depends on the size and polarity of the foreign molecule. Even if a foreign chemical is ingested or touched, only the portion that actually enters the bloodstream results in exposure and risk. Distribution refers to the tissues in the body where a foreign chemical is likely to migrate as it diffuses out of the bloodstream. Elimination is the process by which a foreign chemical is removed from the bloodstream and hence from the body. The two principal organs of elimination are the kidney and the liver, although some foreign chemicals are eliminated by other routes, for example, in exhaled air or in sweat. Uptake, distribution, and elimination are to some degree sequential processes. In particular, uptake must occur before a chemical can be distributed and eliminated. However, there is considerable overlap in time among the three processes.

7. The volume of distribution, or V_d, is a virtual volume. It is proportional to the inverse of the concentration of a bioavailable foreign chemical in blood: The lower the concentration in blood, the larger is the volume of distribution. The V_d is typically referenced to 70-kg males but varies considerably, depending on body weight and fat. In Figure 6.8, extrapolating the shallow portion of the blood concentration-time curve to its intersection with the y-axis where time t = 0 gives a serum (serum is similar to plasma) concentration of 1.07 μg/L (distance of 1.03 log units as measured with a ruler; $10^{1.031}$ = 1.07 μg/L). $V_d = A_b/C_p$ = 25,000 μg/1.07 μg/L = 23.4 L. The literature value is 21 L.

8. Substituting in Equation (6.5): $t_{1/2}$ = 0.693/k = 0.693/3.9 day^{-1} = 0.18 day for chemical Y; $t_{1/2}$ = 0.693/0.01 day^{-1} = 69.3 days for chemical W.

9a. Substituting in Equation (6.8): Amount in body = [(Dose/Dosing interval) × Bioavailability] × 1.44 × $t_{1/2}$ = [(0.5 mg/2.33 days) × 0.15] × 1.44 × 0.18 day = 0.0083 mg. Time to steady state = 3.3 × $t_{1/2}$ = 3.3 × 0.18 day = 0.59 day. The amount in the body fluctuates; however, the average amount reaches 90% of its plateau value after 3.3 half-lives or 0.59 day.

b. Amount in body = [(0.05 mg/0.14 day) × 0.05] × 1.44 × 69.3 days = 1.78 mg. Time to steady state = 3.3 × $t_{1/2}$ = 3.3 × 69.3 = 229 days. Despite the lower dose, longer dosing interval, and lower bioavailability, chemical W builds up to a much higher level in the body than chemical Y due to its longer half-life (slower rate of elimination). Chemical W's longer half-life also means that it takes longer to reach a steady-state level than chemical Y.

REFERENCES

Hodgson, E., and P. E. Levi. 1987. *Modern toxicology*. New York: Elsevier.
Katzung, B. G. 1992. *Basic and clinical pharmacology*. 5th ed. Norwalk, CT: Appleton & Lange.
Shier, D., J. Butler, and R. Lewis. 1996. *Hole's human anatomy and physiology*. 7th ed. Dubuque, IA: Wm. C. Brown.

7 Mechanisms of Chemical Disease

7.1 INTRODUCTION

Disease is a cellular process in which cells malfunction individually and collectively, undermining the tissues and organs they comprise. Toxic chemicals may act in a multitude of ways to attack cells and undermine their function in the body. Toxic effects are generally classified as either immediate (acute) or long term (chronic) (Chapter 5). Acute toxicity is manifested minutes to days following exposure. Examples are carbon monoxide poisoning, snake bites, and drug overdoses. Chronic toxicity is slow in onset, and months or years may pass before a harmful effect becomes apparent. Examples are cancer, infertility, IQ deficit, and hypertension.

In this chapter, molecular and cellular mechanisms of chemical toxicity are organized around the two categories used by the Environmental Protection Agency (EPA) to assess human health risk: noncancer health effects and cancer (Chapter 8). Though simplistic, the EPA categories are useful, because broadly similar principles apply to the actions of all harmful chemicals: A chemical interacts with essential cellular molecules such as DNA or protein, damaging a single cell or a population of cells; the damaged cells are not able to repair themselves or, alternatively, to self-destruct in an orderly fashion; and finally, the damaged cells malfunction, harming the tissue of which they are a part and potentially threatening the life of the organism. Cancer follows the same general pattern as other forms of chemical toxicity except that the injury manifests itself as uncontrolled cell proliferation instead of decline and death.

7.2 NONCANCER HEALTH EFFECTS

Noncancer health effects of toxic chemicals may be divided broadly into organ toxicity and developmental toxicity. Organ toxicity refers to diseases of specific organs or tissues such as liver or blood that result from exposure to toxic chemicals. Developmental toxicity encompasses all forms of damage that result from chemical exposure at critical stages of development, from conception through adolescence to adulthood, and that prevent the organism from reaching its full biological potential. Developmental toxicity encompasses reproductive toxicity, including damage to the reproductive systems of parents and their offspring as well as teratogenesis (birth defects). Fetal alcohol syndrome is a familiar and tragic example of teratogenesis.

7.2.1 Organ Toxicity

Between the time of exposure and the onset of its harmful effect, every toxic chemical progresses through four general stages (Figure 7.1). In the first stage, the toxicant is delivered to the organ or tissue where it will act. In the second stage, the toxicant attacks one or more target molecules; alternatively, it may act by altering the cellular microenvironment. In stage three, the impact of the toxicant on a target molecule or a tissue microenvironment results in injury to cells; unable to repair themselves, cells become dysfunctional, tissue integrity is compromised, and the toxic effect is manifested. The fourth stage, which includes cancer, represents a second category of cellular toxicity whereby mechanisms designed to repair cells are compromised.

The first stage of toxicity, delivery of the toxicant from the site of exposure to the site of action, has already been described (Chapter 6). To recapitulate, exposure begins when a biological receptor, e.g., a person, ingests, inhales, or has skin contact with a toxic chemical. While a few toxic chemicals act locally to damage the gastrointestinal tract, the lungs, or the skin, most must traverse membrane barriers and run a gauntlet of biotransformation reactions before they can reach the heart and be pumped out into the systemic circulation and harm cells. Small, nonpolar, lipophilic molecules are generally able to pass through membranes more easily than polar or charged molecules, and thus they have a better chance of reaching the systemic circulation; however, their progress tends to be impeded by biotransformation enzymes that carry out detoxication reactions, attaching bulky, polar residues that lower foreign chemicals' toxicity and make it easier for them to be excreted in urine or bile. Occasionally, a biotransformation reaction increases rather than decreases the toxicity of a foreign chemical; when this happens, it is called a toxication reaction. The processes that promote and retard toxicant delivery to the target site are summarized in Figure 7.2.

It should be noted that while some toxicants may be said to "seek out" protein molecules on the basis of structurally specific interactions, most molecules become "targets" more or less by accident, e.g., by happening to be in proximity to toxic chemicals as they are distributed in the body and/or as they undergo toxication reactions. They become "targets" mainly by virtue of possessing chemical properties that happen to favor reactions with electrophiles and free radicals, reactions that are relatively nonspecific from a structural standpoint, as discussed later in this section. Indeed, many toxic chemicals react with more than one cellular molecule, and some of their reactions may be toxicologically neutral. Toxicologically neutral reactions may serve a protective function by "soaking up" toxicant molecules that could otherwise engage in chemical reactions that are harmful.

Different toxic chemicals have different effects on target (or "target") molecules in cells:

1. Some toxicants bind noncovalently to proteins such as a receptor, an enzyme, or a transporter molecule, substituting for the naturally occurring small molecule, called a ligand, that normally binds to that site. Some faux ligands block their target protein's function, while others stimulate it, depending on the particulars of the chemical interaction. Examples of

Mechanisms of Chemical Disease

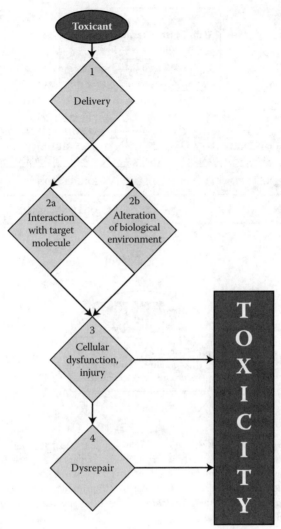

FIGURE 7.1 General stages in the internal development of disease following exposure to a toxic chemical. (Reprinted with permission from Curtis Klaassen, ed., *Casarett and Doull's Toxicology, The Basic Science of Poisons*, 6th ed. [New York: McGraw-Hill, 2001], 36.)

noncovalent binding are carbon monoxide, a structural analog of oxygen that inhibits oxygen uptake by binding to hemoglobin in its place; tubocurarine, which blocks the binding of acetylcholine in the neuromuscular junction in skeletal muscle, causing muscle relaxation and, at higher doses, general paralysis and suffocation (Europeans learned about curare from South American Indians, who used it on poison darts); and structural analogs of hormones, e.g., estrogen mimics, which may counter or promote the actions of the endogenous hormone. Noncovalent interactions, though structurally specific, usually involve weak chemical bonds between toxicant and target molecules. Consequently, toxic effects are generally reversible by

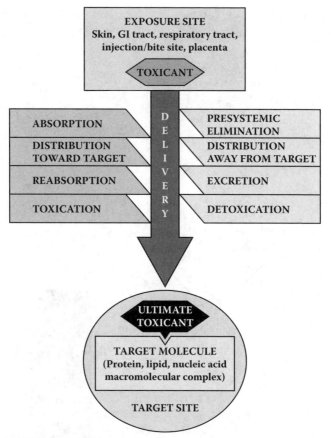

FIGURE 7.2 Overview of physiological processes that affect the delivery of a toxicant molecule from the point on the body where exposure occurs to its target site in the body. The processes on the left promote delivery of toxicants, while the processes on the right hinder toxicant delivery to target sites (Chapter 6). (Reprinted with permission from Curtis Klaassen, ed., *Casarett and Doull's Toxicology, The Basic Science of Poisons*, 6th ed. [New York: McGraw-Hill, 2001], 38.)

lowering the concentration of toxicant in the body—for example, by moving an unconscious victim from a house filled with carbon monoxide fumes to fresh air outdoors. As the victim is removed from exposure, the body's concentration of carbon monoxide (or other noncovalently binding toxicant) decreases, because toxicant molecules dissociate from their target molecules according to the law of mass action and are excreted—in the victim's breath in cases of gaseous toxicants like carbon monoxide, or in urine in cases of dissolved toxicants like curariform drugs and hormone analogs.

2. Some toxicants bind covalently to target molecules such as protein or DNA. There are several classes of toxicant molecules that undergo covalent binding, the largest of which are the electrophiles. Electrophiles commonly result from toxication reactions whereby biotransformation enzymes, which evolved to protect the body from foreign chemicals,

are chemically "duped" into endangering it instead. Electrophiles created by toxication reactions are molecules with an overall deficiency in electrons. Some electrophiles carry a positive electrical charge, while many do not. Electrophiles home in on electron-rich (also called nucleophilic) atoms in protein and DNA molecules, forming covalent bonds that are strong and, with few exceptions, irreversible. A toxic chemical that binds covalently to a protein or DNA molecule is called an adduct. Examples of adducts are the covalent binding of aflatoxin to guanine residues in DNA, which can initiate the process of carcinogenesis; and the covalent binding of malathion and other organophosphate insecticides to the active site of the enzyme acetylcholinesterase, which causes a rapid buildup of the neurotransmitter acetylcholine, resulting in paralysis and death.

3. Occasionally the covalent binding of a toxic chemical adduct changes the structure of a cell surface protein molecule (Figure 6.3) in such a way that it is no longer recognized as "self" and is attacked by the body's immune system. The altered protein is called a neoantigen ("new antigen," i.e., a newly formed antigen; an antigen is any protein molecule perceived as nonself by the immune system). Two examples of neoantigen formation that result in autoimmune diseases are drug-induced lupus erythematosus and sensitivity to the general anesthetic halothane, which produces a hepatitis-like syndrome in some patients.

4. Some toxicants participate in chemical reactions inside the cell that promote the formation of free radicals of oxygen and nitrogen. Free radicals contain one or more unpaired electrons that make them highly reactive. Free radicals are capable of destroying cellular molecules. For example, the hydroxyl radical can initiate a sequence of chemical reactions in lipids, called lipid peroxidation, which result in the fragmentation of lipid molecules in cell membranes, undermining membrane integrity and function. The hydroxyl radical can also cause breaks in DNA molecules.

Free radicals and electrophiles, which act by forming covalent bonds, tend to be less specific with respect to the molecules they attack than noncovalently binding toxicants, which act by substituting for specific ligands on proteins. Nevertheless, whether the chemical interaction involves a covalent or a noncovalent bond, the overall effect is to undermine some aspect of cells' ability to maintain internal functions that are critical to their survival. A prominent example of a critical internal function is energy metabolism. Many toxicants act on molecules that are required for the synthesis of ATP (adenosine triphosphate). Called the "energy currency of the cell," ATP is indispensable as the cell's major source of energy; it is also an integral component of cellular metabolism and takes part in the biosynthesis of a wide range of essential molecules. ATP is generated continuously in a complex set of reactions inside mitochondria. A wide range of toxic chemicals undermine the mitochondrial synthesis of ATP. Examples are cyanide, which binds noncovalently to cytochrome oxidase and inhibits electron transport; salicylate (aspirin), which dissipates the electrical potential across the mitochondrial membrane and thereby impedes the final step of ATP synthesis; and carbon monoxide, one of a number of toxicants that undermine ATP synthesis by inhibiting the delivery of oxygen to mitochondria.

Another cellular function that is undermined by toxic chemicals is the maintenance of low intracellular concentrations of calcium. The calcium concentration inside cells is 10,000 times lower than the calcium concentration in extracellular fluid and blood. Increased intracellular calcium is potentially lethal to cells because it has multiple effects, including inhibition of ATP synthesis, degradation of the cytoskeleton responsible for maintaining cell shape, and activation of proteolytic enzymes that then degrade essential cellular proteins. Cells maintain their low internal calcium concentration by two kinds of mechanisms: low membrane permeability to calcium, and internal calcium "pumps" that continuously extrude calcium out of the cell cytoplasm. Some toxicants act by promoting the influx of calcium, while others act by inhibiting its efflux. For example, methylmercury induces the formation of porelike openings in the cell membrane that allow calcium to move down its concentration gradient into the cell. Some snake venoms contain enzymes that act by "chewing up" cell membranes and letting calcium in. The efflux of calcium out of cells is inhibited by toxicants that bind to and impair the cell's internal calcium pumps, for example, chloroform and cadmium. The general types of harmful interactions between toxic chemicals and cellular target sites are summarized in Figure 7.3.

7.2.2 Developmental Toxicity

The developing organism passes through a number of stages of rapid change where it is singularly vulnerable to assault by specific toxicants on particular sets of cells involved in its growth. Developmental toxicity, which encompasses reproductive toxicity, can combine features of acute and chronic disease when the period of exposure is brief, yet the toxic effect lasts a lifetime. Some marquee examples of developmental toxicity have already been mentioned: fetal alcohol syndrome, the thalidomide tragedy in Europe in the 1960s, and intersex fish. In addition to brief exposures at critical stages of development, there is also concern that ongoing exposure to low doses of toxicants could harm slower developmental processes that take months or years to unfold. For example, circumstantial evidence implicates environmental estrogens as one of several factors in the falling age of puberty in U.S. girls. The heightened vulnerability of developing organisms to chemical stressors is becoming an increasingly important consideration in the management of toxic chemical risk.

7.3 CANCER

Cancer is arguably the best-studied form of chronic chemical toxicity. It is manifested by the growth of a tumor, which typically appears years after exposure. A tumor is in essence a colony of mutant cells that crowds out normal cells and eventually destroys the tissue it inhabits. In some forms of cancer called metastatic, the tumor cells migrate to other organs and destroy them, too. The term "carcinogenesis" refers to the years-long process by which cancer unfolds.

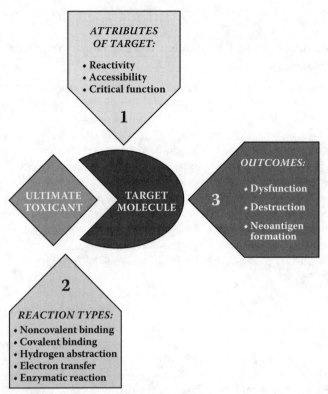

FIGURE 7.3 Interactions of toxicant molecules with target molecules in cells. In addition to noncovalent and covalent binding, interactions with toxicant molecules include the removal of protons as well as redox reactions involving the transfer of electrons. Rarely, a toxicant is an enzyme that acts by attacking and degrading a target molecule directly, e.g., toxins in some snake venoms. The term "ultimate toxicant" refers to the final form of a toxic chemical molecule when it reaches the target molecule, including any structural alterations resulting from toxication reactions. (Reprinted with permission from Curtis Klaassen, ed., *Casarett and Doull's Toxicology, The Basic Science of Poisons*, 6th ed. [New York: McGraw-Hill, 2001], 44.)

A variety of agents in the human environment have been linked to cancer over the past two centuries (Table 7.1). Often, the link between environmental exposure and cancer is first suggested by epidemiological observations of at-risk populations (Chapter 4); however, decades often pass before scientific evidence confirms the link. In a particularly noteworthy example of the lag between epidemiological evidence and conclusive proof of causation, Percival Pott, a physician in London, postulated in 1775 that soot and coal tar cause cancer of the scrotum, because in his medical practice he observed that many of the patients who presented with scrotal cancer were chimney sweeps. Yet it was not until 150 years later in the 1920s that polycyclic aromatic hydrocarbons (PAHs) were isolated in the laboratory and shown in animal studies to be the active carcinogenic ingredients in coal tar, e.g., 3-methylcholanthrene and 7,12-dimethylbenz(a)anthracene (Figure 7.4). Many naturally occurring chemicals also cause cancer, acting by the same general mechanisms as industrial chemicals (Figure 7.5).

TABLE 7.1
Years When Agents in the Environment Were Originally Proposed to Cause Cancer in Humans

Suspected Carcinogen	Organ	Discoverer	Year
Soot	Scrotum	Pott	1775
Pipe smoking	Lips	Sommering	1795
Coal tar	Skin	Volkman	1875
Dye intermediates	Bladder	Rehn	1895
X-rays	Skin	Van Trieben	1902
Tobacco juices	Oral cavity	Abbe	1915
Radioactive watch dial dyes	Bone	Martland	1929
Sunlight	Skin	Molesworth	1937
Tobacco (cigarette) smoking	Lung	Muller	1939
Asbestos	Pleura	Wagner	1960
Cadmium	Prostate	Kipling; Waterhouse	1967

Source: Reprinted with permission from Ernest Hodgson and Patricia Levi, *Modern Toxicology* (New York: Elsevier, 1987), 146.

Note: Proposals were based on observations of at-risk human populations, e.g., chimney sweeps in London. Decades typically passed before definitive proof of cancer causation was provided by laboratory studies. In the past half century there have been increased efforts to identify carcinogenic agents proactively by means of chronic toxicity testing in rodents (Chapter 5).

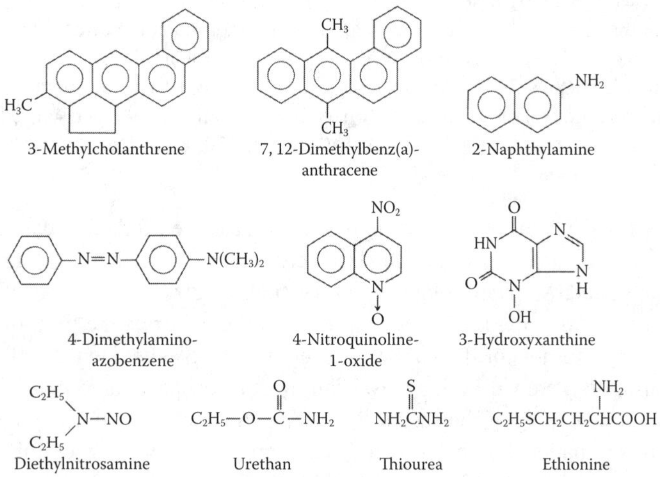

3-Methylcholanthrene   7,12-Dimethylbenz(a)-   2-Naphthylamine
                            anthracene

4-Dimethylamino-   4-Nitroquinoline-   3-Hydroxyxanthine
azobenzene   1-oxide

Diethylnitrosamine   Urethan   Thiourea   Ethionine

FIGURE 7.4 Structures of selected industrial chemicals that have been shown to cause cancer. Many carcinogenic chemicals interact with DNA, causing mutations that are necessary but not sufficient for cancer to develop. (Reprinted with permission from Ernest Hodgson and Patricia Levi, *Textbook of Modern Toxicology* [New York: Elsevier, 1987], 146.)

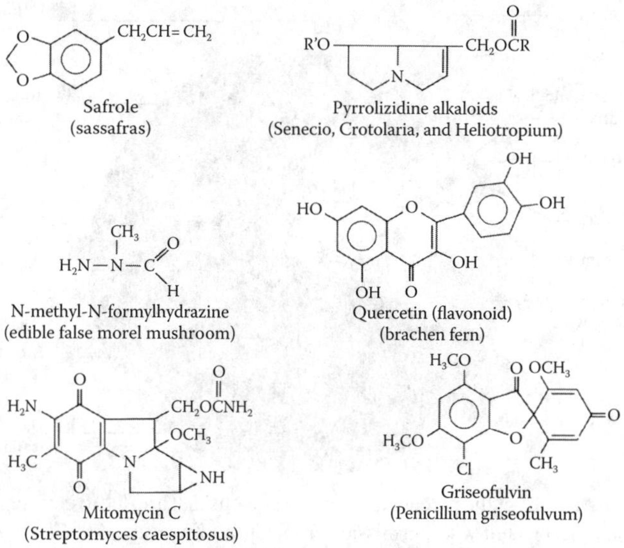

Mechanisms of Chemical Disease

FIGURE 7.5 Structures of selected naturally occurring chemicals that have been shown to cause cancer. Naturally occurring chemicals cause cancer by the same general mechanisms as industrial chemicals. (Reprinted with permission from Ernest Hodgson and Patricia Levi, *Textbook of Modern Toxicology* [New York: Elsevier, 1987], 147.)

Cancer can and does occur spontaneously as a result either of errors in the intricate processes by which the genetic material, DNA, is replicated and maintained in the cell or as a result of exposure to naturally occurring chemicals and radiation in the environment. Cancer in excess of background levels is related to a host of factors involving lifestyle, occupation, and geography (Figure 7.6). It is interesting to note that tobacco, diet, and infections are thought to account for a large majority of cancer deaths.

Cancerous tumors, whether arising from spontaneous errors in DNA replication and maintenance or from exposure to carcinogenic chemicals, either naturally occurring or human-made, are thought to develop in three broad stages: initiation, promotion, and progression.

1. Initiation begins with damage to the structure of a cell's DNA. Naturally occurring as well as industrial chemicals can damage DNA and initiate the process of carcinogenesis (Figures 7.4 and 7.5). DNA structural damage is necessary but not sufficient to initiate carcinogenesis. Cells have been equipped by evolution to repair most types of damage, provided the DNA alterations are not too large or too numerous. Repairs are carried out by specialized enzymes. Some repair enzymes snip out damaged pieces of DNA and replace them with intact pieces. Other enzymes carry out their repairs when DNA unwinds and each strand replicates itself in preparation for cell division; these enzymes are able to modify the DNA replication

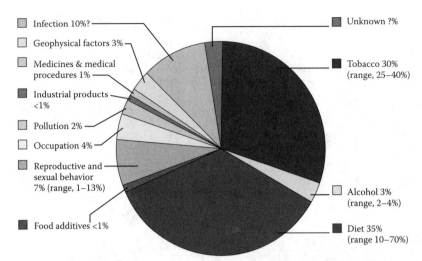

FIGURE 7.6 Estimates of the percentage of cancer deaths that are due to environmental factors of various kinds. (Reprinted with permission from Curtis Klaassen, ed., *Casarett and Doull's Toxicology, The Basic Science of Poisons*, 6th ed. [New York: McGraw-Hill, 2001], 280.)

process in ways that restore the integrity of the DNA molecule. In the vast majority of cases, the damage is repaired and cells are restored to normalcy without any harm to the organism (Figure 7.7).

If a cell fails in its efforts to repair its DNA, it may die. Alternatively, the damage becomes permanent as an irreversible mutation that persists in future generations of the cell line and has the potential to develop into a tumor. Like the original DNA damage, however, a mutation is also not sufficient to cause cancer. Most mutations are "silent," meaning they are not expressed by the cell (Figure 7.7). A mutation may be silent because it occurs in a part of the DNA that is not critical to the cell's health. Alternatively, a mutation may have the potential to give rise to a cancerous cell, but the potential is never realized. In order for a cancerous mutation to be expressed by a cell, the environment inside the cell has to favor its expression. Cellular processes that collectively favor the expression of a mutation's carcinogenic potential are called promotion.

2. Promotion refers to a set of cellular processes by which the normal balance between orderly cell replication and orderly cell death is tilted in favor of cell replication, ultimately resulting in the unrestrained replication of a cell line to form a tumor. The pivotal role of promotion in carcinogenesis shows that cancer is first and foremost a disease of the regulation of cell growth. Some examples of chemicals that promote carcinogenesis are shown in Figure 7.8.

Virtually all of the body's cells (except nerve cells) divide continuously throughout life. Cells divide at different rates in different tissues and at different stages of an organism's life, but in general, new cells are produced and old cells die continuously throughout the life cycle. In a healthy organism, old cells self-destruct in an orderly sequence of events called apoptosis

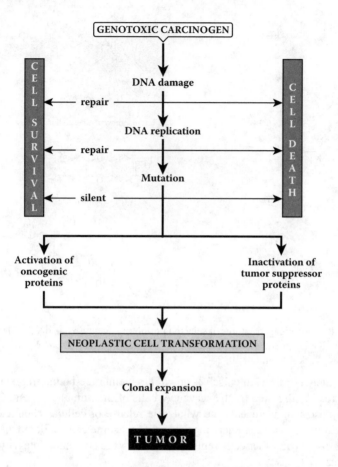

FIGURE 7.7 The three stages of carcinogenesis set in motion by exposure to an external agent, or carcinogen. In the first stage, initiation, the cell's DNA is damaged. If the cell's repair mechanisms fail to correct the damage, and if the cell does not die, a mutation results. No harm comes to the cell if the mutation is silent. If the mutation is not silent, then the second stage of carcinogenesis, promotion, determines whether the cell is transformed from a normal cell to a tumor cell. This transformation depends on the balance between oncogenic proteins and tumor suppressor proteins. If oncogenic activity gains the upper hand, the cell loses the growth constraints of normal cells and starts to replicate continuously. In the third and final stage of carcinogenesis, the clone of continuously replicating cells forms a visible tumor that crowds out healthy cells and eventually destroys the tissue it inhabits. Cancer need not be initiated by a mutation in DNA. It can also be caused by chemicals called nongenotoxic carcinogens that act by promoting cell replication. See text for details. (Reprinted with permission from Curtis Klaassen, ed., *Casarett and Doull's Toxicology, The Basic Science of Poisons*, 6th ed. [New York: McGraw-Hill, 2001], 73.)

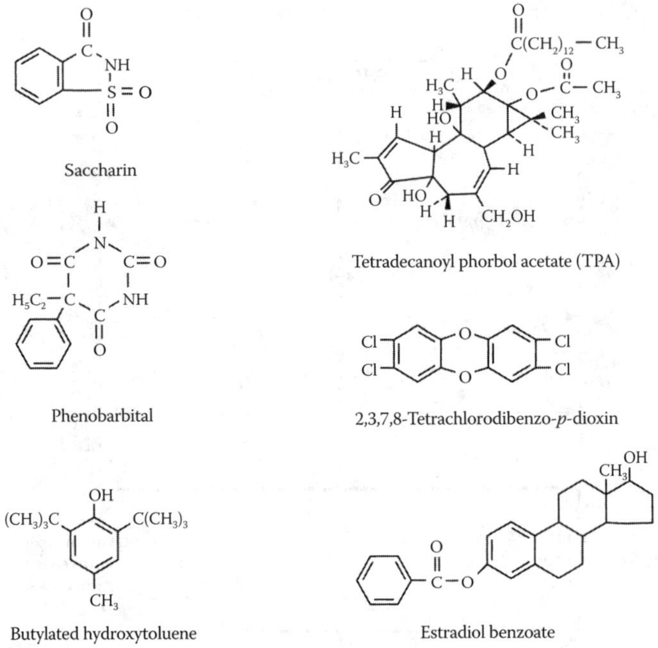

FIGURE 7.8 Chemical structures of selected promoting agents. (Reprinted with permission from Curtis Klaassen ed., *Casarett and Doull's Toxicology, The Basic Science of Poisons*, 6th ed. [New York: McGraw-Hill, 2001], 268.)

in which they break themselves down into membrane-bound fragments that are readily digested by the scavenger cells of the immune system. Orderly self-destruction prevents the wholesale release of cellular chemicals that might otherwise harm neighboring cells. As some cells undergo apoptosis and disappear, other cells replicate, ensuring the continued integrity of the tissue.

Orderly cell replication and orderly cell death are regulated by different groups of genes. For cell replication, two sets of genes are recognized as critical: the so-called proto-oncogenes and the tumor suppressor genes. The proto-oncogene family includes genes coding for proteins that promote cell replication, e.g., growth factors, signal transducers, and promoters of DNA transcriptional activity. A well-understood example of a proto-oncogene are the members of a group of signal-transducing proteins that are coded by genes called Ras. The proteins coded by the Ras gene family are located on the inside of the cell membrane, where they play a crucial role in relaying signals from growth factors that bind to the surface of cells to proteins in the cytoplasm and nucleus, signals that have the effect of stimulating cells to divide.

While proto-oncogenes foster cell replication, a second group of genes suppresses replication. Dubbed tumor suppressor genes, they code for proteins that counter the growth-promoting actions of the proto-oncogenes. Tellingly, tumor suppressor genes also promote apoptosis. Apoptosis may serve as a defense against tumors because it can rid the body of a mutated cell before it has a chance to develop into a tumor. Thus tumor suppressor genes appear to perform a dual function of maintaining cell replication at a pace consistent with the needs of the tissue and weeding out precancerous cells by ushering them into apoptosis. The best-known tumor suppressor gene, called p53, is commonly referred to as the "guardian of the cell."

Cancer develops when, simply put, a cell escapes the control of the tumor suppressor genes, and the balance between cell replication and cell death shifts in favor of replication. Often this shift is caused by mutations in proto-oncogenes themselves, mutations that allow them to evade the control of tumor suppressor genes. The shift may also result from mutations in tumor suppressor genes that impair their ability to keep proto-oncogenes in check. Mutations in genes other than proto-oncogenes and tumor suppressor genes may also lead to cancer. Further, even in the absence of an initiating genetic mutation, exposure to growth-promoting chemicals in the environment may be sufficient for some forms of cancer to develop; the promoting chemical then acts as a so-called nongenotoxic carcinogen. For example, estrogenlike industrial chemicals such as polychlorinated biphenyls (PCBs) and the pesticides DDT and atrazine bind to estrogen receptors on the surface of cells and are thought by some scientists to be capable of promoting cancer by stimulating excessive cell replication in breast and other estrogen-responsive tissues.

One of the most significant features of the promotion stage of carcinogenesis is that promotion is reversible, whereas the initiation and progression stages are essentially irreversible processes. Thus, it is entirely possible for a cell or a population of cells to maintain their normal rate of replication and apoptosis despite a mutational change in their DNA and/or exposure to chemicals in the environment that promote cell replication. A tumor develops when, in the promotion stage, cells' ability to maintain their growth-death equilibrium is overwhelmed.

3. Progression, the final stage of carcinogenesis, is characterized by the irreversible commitment of a population of cells to multiply faster than needed to maintain the healthy functioning of a tissue. Cells in the initiation and promotion stages of carcinogenesis cannot be distinguished from healthy cells on the basis of their appearance under the microscope. In the progression stage, however, the rapidly growing cell population develops into a visible tumor, also referred to as a neoplasm (Figure 7.7). The neoplasm may be benign or malignant. If it is malignant, it may be confined to its tissue of origin, or it may metastasize to other tissues.

The DNA in tumor cells tends to become increasingly unstable. DNA instability may manifest itself in changes in the appearance of a cell's chromosomes when they condense immediately prior to cell division, or mitosis. The appearance of the complete set of condensed chromosomes at mitosis is called the karyotype. Changes in tumor cells' karyotypes may include breaks or rearrangements within chromosomes or the translocation of a piece of one chromosome to another chromosome. DNA instability increases the risk of fresh mutations that may further accelerate a tumor's growth.

Medical treatments include surgery to remove the tumor as well as chemotherapy and radiation to kill the cancerous cells. Chemotherapy turns cancer cells' rapid multiplication against them by targeting proteins that are essential for cell division. Unfortunately, since healthy cells also divide, they too are impacted, and the side effects of chemotherapy can be severe.

Chapters 6 and 7 have introduced the major features of toxic chemicals' disposition and effects inside the body, including membrane barriers to their penetration into the systemic circulation, biotransformation by detoxication and toxication reactions, and types of interactions with target molecules and the forms of damage done to cells. In the following chapters, we return to the view of toxic chemical risk from outside the body. Chapter 8 describes the assessment of cancer and noncancer risks to human health, and Chapter 9 introduces the discipline of ecological risk assessment, a complex yet elegant approach to understanding and managing risks of toxic chemicals to natural systems.

STUDY QUESTIONS

1. Describe the major kinds of biological molecules that are "targeted" by toxicant molecules and their locations in the cell.
2. True or false: Covalent toxicant interactions with "target" molecules are generally more specific than noncovalent interactions. Explain.
3. Discuss the attributes of biological molecules that make them the "targets" of toxic chemical molecules.
4. Name three specific toxicants and/or classes of toxicants that act by binding noncovalently to "target" molecules in the cell.
5. Describe two major chemical classes of toxicants that act by forming covalent chemical bonds with "target molecules." Name one specific toxicant-target interaction for each class.
6. Identify three specific toxicants that act by interfering with ATP synthesis in mitochondria. Why is ATP synthesis essential to cell function?
7. Briefly describe the three stages of carcinogenesis. Does the interaction of a toxicant molecule with DNA lead inevitably to cancer? Explain.
8. What are the two sets of genes that regulate cell growth (i.e., cell replication)? Name one example from each set of genes.
9. Can cancer develop in the absence of a genetic mutation? Explain.
10. What is apoptosis? How might it help defend the body against cancer?

ANSWERS TO STUDY QUESTIONS

1. There are three major classes of cellular molecules that are "targeted" by toxicant molecules: proteins, DNA, and lipids. Proteins perform a vast array of functions and are located throughout the cell. DNA is found mostly in the cell nucleus; there is also a small but significant amount of DNA in mitochondria. Lipids are concentrated in membranes such as the outer membrane enclosing the cell, the cell's internal network of membranes

called the endoplasmic reticulum, and mitochondrial membranes (see also Figure 6.1).
2. False. Noncovalent interactions tend to be much more specific than covalent interactions from the standpoint of the structural complementarity between toxicant and target molecules. In addition to being structurally more specific, noncovalent interactions are also generally weaker than covalent interactions. That noncovalent bonds tend to be both structurally more specific and chemically weaker than covalent bonds appears paradoxical unless it is recognized that structure and bonding are distinct (though overlapping) aspects of molecular pairing: Structure determines which molecules recognize and interact with one another, while the nature of the chemical bond determines how strong their interaction is. Structural specificity makes it possible for a noncovalently binding toxicant to "seek out" a particular class of target molecules in the body while the weakness of the noncovalent bond means its toxic action persists only as long as its concentration in the cell remains high enough to drive the noncovalent association with its target molecule (law of mass action, Chapter 3).
3. In the case of noncovalent interactions, target molecules are characterized by their structural complementarity with toxicant molecules. Toxicants are often structurally similar to endogenous ligands, i.e., molecules produced naturally by the body for specific purposes, for example, hormones. In the case of covalent interactions, biological molecules frequently become targets more or less by accident, e.g., they happen to be in the vicinity when a toxicant molecule is taken up into the bloodstream or when it is formed by a toxication reaction in the cell. Some covalently binding toxicants interact preferentially with certain types of sites on biological molecules, e.g., the carcinogen aflatoxin preferentially attacks guanine residues in DNA. In general, a covalently binding toxicant attacks sites that allow it to form its particular brand of covalent bond, regardless of where in a large protein, lipid, or DNA molecule the chemically compatible site happens to be located. Noncovalently binding toxicants require structural compatibility analogous to a key fitting in a lock.
4. Three noncovalent toxicants are carbon monoxide, which displaces oxygen from its binding site on hemoglobin; tubocurarine, which displaces the neurotransmitter acetylcholine from its nicotinic receptor in the neuromuscular junction and is used in anesthesia to relax skeletal muscle; and estrogen analogs, which bind to estrogen receptors and which may stimulate or block their activity, depending on each analog's particular structural "fit" with the receptor.
5a. Electrophiles. Example: Formation of an aflatoxin adduct with guanine residues of DNA.
b. Free radicals of oxygen. Example: Lipid peroxidation initiated by the hydroxyl free radical.
6. Three toxicants that inhibit ATP synthesis by mitochondria are cyanide, aspirin, and carbon monoxide. ATP is the "energy currency" of the cell;

without it, all cells soon die. It is also an essential component of many biosynthetic reactions that replenish cellular molecules.
7. The three stages of carcinogenesis are initiation, promotion, and progression. Initiation typically begins with a mutation in a cell's DNA that creates a potential for cancer. Promotion refers to a complex set of cellular processes that are affected by the mutation such that the cell line's replication-apoptosis equilibrium is skewed toward replication and ultimately eventuates in uncontrolled replication. Progression is the growth of the colony of neoplastic cells into a visible tumor. Toxicant interaction with DNA does not inevitably result in cancer. For example, the cell may repair the damaged DNA. If DNA repair fails and a mutation results, the mutation may be silent. If the mutation is expressed, its action may be suppressed in the promotion stage.
8. The two broad classes of genes that code for proteins that regulate cell growth are the proto-oncogenes and the tumor suppressor genes. The Ras genes are a family of proto-oncogenes that are involved in cell signaling and promote cell replication. The p53 gene slows the pace of cell replication and encourages apoptosis.
9. Yes. Promoting agents can act as nongenotoxic carcinogens in some cases.
10. Apoptosis is an orderly process by which aging cells self-destruct, making way for new cells generated by replication. Apoptosis may serve as a defense against cancer by eliminating mutated, precancerous cells during the promotion stage of carcinogenesis.

REFERENCES

Hodgson, E., and P. E. Levi. 1987. *Modern toxicology*. New York: Elsevier.
Klaassen, C. D., ed. 2001. *Casarett and Doull's toxicology: The basic science of poisons*. 6th ed. New York: McGraw-Hill.

SUGGESTED READING

Steingraber, S. 2007. *The falling age of puberty in U.S. girls*. San Francisco: Breast Cancer Fund.

8 Human Health Risk Assessment

8.1 INTRODUCTION

Risk assessment is designed as a science-based tool to help policy makers discern specific public health threats, to gauge their relative severity, and to establish priorities and allocate limited resources to address them. Used appropriately, risk assessment serves as a rational basis for protecting human health and the environment from risks due to toxic chemicals. However, risk assessment is not a panacea. Scientific uncertainty is a fundamental characteristic of all risk assessments, and uncertainty places limits on risk assessment as a decision-making tool. How to prioritize risks that are judged to be significant but whose relative magnitudes are difficult to gauge represents a considerable challenge. Nevertheless, risk assessment provides a basis for stakeholders to exercise reason, imagination, and common sense as they go about the arduous task of negotiating acceptable levels of risk from toxic chemicals.

It is important not to confuse risk assessment with risk perception. Risk perception refers to an emotional response that makes a person more willing to accept some types of risk than others. One key to acceptance is the perception that a risk can be observed and/or controlled. Examples are smoking cigarettes and driving a car, both of which, from the standpoint of risk assessment, rank among the riskiest activities of contemporary life. While familiar risks may be perceived as less serious than they actually are, risks that are invisible or involuntary or that can affect future generations are sometimes attended by fear and dread, for example, nuclear power plants, radioactive waste, and industrial chemicals that cause cancer—even though, from the standpoint of expert risk assessment, the probability that they will cause harm to an individual is relatively low. Differences between lay and expert rankings of risk are suggested by the results of a survey reported in Table 8.1.

While public perception of risk is sometimes frustrating to expert risk assessors, it can serve a useful purpose by focusing attention on catastrophic risks that impact whole communities, not just individuals. The risk of a meltdown at a nuclear power plant is small, but the consequences are huge: Witness Chernobyl and the continuing effects of radioactive contamination. In the difficult political business of managing risks, risk perception can complement expert risk assessment and help drive the regulatory process forward.

TABLE 8.1
Selected Examples of Differences Between Expert and Lay Perceptions of Risk

Activity/Agent	Expert Rank	Lay Rank
Motor vehicles	1	2
Smoking	2	4
Alcoholic beverages	3	6
Handguns	4	3
Surgery	5	10
Motorcycles	6	5
X-rays	7	22
Pesticides	8	9
Electric power (nonnuclear)	9	18
Swimming	10	19

Source: Reprinted with permission from Mary Amdur, John Doull and Curtis Klaasen, eds., *Casarett & Doull's Toxicology, The Basic Science of Poisons*, 4th ed. (New York: Pergamon Press, 1991), 986.

Note: Nuclear power was ranked 1 on the lay list and 20 by experts. The rankings are from 1979.

8.2 THE PROCESS OF RISK ASSESSMENT

A risk assessment is performed in four steps: hazard identification, analysis of exposure, analysis of effect, and risk characterization. The same general process is used to assess risk from many different types of threats, not just toxic chemicals. Risk assessments are performed for individual chemicals. When exposure is to a mixture of chemicals, risk assessments are performed for each individual chemical in the mixture. Generally, the health risks from individual chemicals are added together to estimate the total health risk from the mixture of chemicals. In other words, health risks are generally considered to be additive. If there is evidence that two chemicals either enhance or interfere with each other's toxicity, then that information is factored into the risk assessment, usually in the risk characterization step.

8.3 HAZARD IDENTIFICATION

The chemical risk-assessment process begins by identifying chemicals of concern, potential pathways of exposure, and at-risk populations. Consider, for example, a Superfund hazardous chemical waste site, a pesticide manufacturing facility that operated from 1950 to 1970, when it went out of business, and was subsequently razed. Its 20 years of operation were characterized by sloppy waste disposal practices and numerous complaints from nearby residents to local authorities. The site was reported to the U.S. Environmental Protection Agency (EPA) in 1983, and a preliminary site investigation was conducted pursuant to the federal Comprehensive Environmental Response, Compensation and Liability Act of 1980 (CERCLA, or

Superfund, see Table 4.1). The preliminary investigation identified numerous chemicals in on-site soil, the groundwater aquifer beneath the site, and the sediment of an unnamed stream that received runoff from the site. The chemicals showing the highest concentrations were arsenic, a common ingredient of older pesticides, and dinoseb, an herbicide. Potential exposure pathways and at-risk populations included: (a) the ingestion of soil by dirt-bikers and others who trespassed on the site; (b) the future use of groundwater as a drinking water source if a residential neighborhood were developed after the Superfund site was cleaned up; and (c) the ingestion of sediment by children who played in the off-site stream. Separate risk assessments needed to be performed for each chemical, exposure pathway, and at-risk population. In the case of arsenic, the most abundant contaminant, three separate risk assessments were required: arsenic in drinking water, arsenic in on-site soil, and arsenic in off-site stream sediment.

As a second example of hazard identification, consider a coal-burning electric power plant, which is regulated under the Clean Air Act. Hazardous substances it emits from its smokestack include carbon dioxide, sulfur dioxide, nitrogen dioxide, particulate matter, and mercury (not a complete list). All of these substances impact human health, including carbon dioxide, which was recently (April 2009) ruled by EPA to have an adverse impact on human health because it is a greenhouse gas that contributes to global warming. Sulfur dioxide is converted to sulfuric acid when it reacts with oxygen and water droplets in the air (Chapter 2), and the acid has adverse impacts on the lungs. Nitrogen dioxide promotes a photochemical reaction resulting in the formation of excess atmospheric ozone, a prominent component of smog, also a known lung stressor. Particulate matter, especially very fine small-diameter particles, also impairs lung function. Mercury causes birth defects and is neurotoxic (Minamata disease, Chapter 2).

Turning to the identification of potential exposure pathways, the three stressors of pulmonary function—sulfuric acid, ozone, and particulate matter—primarily impact populations living downwind from the power plant. The height of the smokestack and weather factors such as the speed and direction of the wind and amount of sunlight (sunlight enhances the photochemical production of ozone) determine the concentration and extent of the pollutant plume on any given day, from a few miles to a hundred miles or more. Every person downwind who breathes air containing these pollutants is at risk. Several subpopulations are at increased risk, for example, those with preexisting pulmonary diseases such as asthma and emphysema, the elderly, and the infirm. Pathways of mercury exposure tend to involve water: When it falls on oceans, lakes, and other water bodies, it is taken up by small organisms such as bacteria and phyto- and zooplankton at the base of aquatic food chains. Some microorganisms can biotransform mercury to methylmercury, an organic compound that is taken up more readily and hence is more toxic than metallic mercury (quicksilver). Mercury and methylmercury undergo bioconcentration and biomagnification. Fish may have mercury levels hundreds or thousands of times higher than algae. Human beings are at the top of these food chains and as such are at greatest risk from mercury. While mercury, sulfuric acid, ozone and particulate matter act directly, carbon dioxide acts indirectly by contributing to increases in average global temperatures. The exposure pathway is in effect the entire atmosphere, and at-risk populations include the entire biosphere. Predicted effects of rising global temperatures include increases in droughts, water shortages, storms, floods, and infectious diseases.

A pesticide-contaminated Superfund site and a coal-burning electric power plant illustrate the general principles involved in hazard identification, the first step in the risk-assessment process: Identify hazardous chemicals to which people might be exposed, the pathways by which exposure might occur, and the populations and subpopulations that are at risk. Once hazards have been identified, the risks they pose can be assessed. Risk is a combination of exposure and toxicity (Figure 1.1). An assessment of risk continues with an analysis of the degrees to which at-risk populations are exposed to chemicals of concern.

8.4 ANALYSIS OF EXPOSURE

Obtaining an accurate exposure estimate for an at-risk population is a difficult task. Indeed, exposure is generally the single greatest source of uncertainty in a risk assessment. Exposure estimates are based on many assumptions, and even small changes in assumptions can result in large differences in estimated risk. The uncertainty surrounding exposure is partly due to the difficulty of predicting "real world" behaviors of at-risk populations that may bring them into more or less contact with a chemical of concern. Nor is it easy to predict how the chemical itself will behave once it is "out in the world," i.e., what its environmental fate and transport will be. Part of the value of an exposure analysis is that it spells out assumptions on which exposure estimates are based.

8.4.1 Chronic Daily Intake

The goal of exposure analysis is to estimate the chronic daily intake (CDI) of a chemical of concern. The chronic daily intake is the amount, or dose, of a chemical that enters the bloodstream of members of the at-risk population averaged over the number of days the population is exposed. For example, consider the ingestion of arsenic in drinking water by residents who move into a neighborhood built on a former Superfund site (see Section 8.3). Assessment of this hazard is based on the possibility that the EPA or state regulatory agency may decide to require that the Superfund site be cleaned up to residential standards and that the aquifer beneath the site be suitable for use as a drinking water source. In order to make a decision about cleanup, the regulatory agency needs a baseline risk assessment. Exposure of future residents by this pathway is the product of three separate factors:

1. The concentration of arsenic in the groundwater aquifer
2. The volume of water a resident drinks
3. The fraction of ingested arsenic that is bioavailable, i.e., that is taken up into the bloodstream from the intestinal tract

Thus, the exposure (dose) due to drinking water from the contaminated aquifer is the product of (arsenic concentration in drinking water aquifer) × (volume of water ingested) × (fraction of arsenic that is bioavailable).

The question is, how can useful numbers be placed on each of these factors? Data on bioavailability may be present in the scientific literature on arsenic. If not, or if studies are inconclusive, bioavailability is assumed conservatively to be equal to 1 (100%). The amount of tap water a person drinks can vary significantly from day to day and from person to person. To promote consistency in risk assessments, EPA has defined a number of default values for ingestion of water and food, inhalation of air, and other routes of exposure (Table 8.2). EPA's default values are generally conservative, meaning they tend to err on the side of overestimating rather than underestimating exposure. The default value for the residential ingestion of potable water is two liters (approximately two quarts, or half a gallon) per day.

The concentration of arsenic in the groundwater aquifer, the third factor in determining the dose of arsenic to future residents through drinking water, is even harder to put a number on than bioavailability or ingestion. The arsenic concentration in the aquifer is estimated by drilling monitoring wells at several locations around the Superfund site, withdrawing groundwater samples, and sending them to a state-certified laboratory for measurement of arsenic concentrations. The concentration of arsenic may vary by a factor of ten, a hundred, a thousand, or more from one monitoring well to another, depending on how much arsenic was dumped at various locations around the property and how its movement into groundwater was influenced by the microenvironment of soil and hydrogeology at the locations where dumping occurred.

To illustrate the kind of variability that is typical of the concentrations of toxic chemicals measured in water—and in environmental media in other risk

TABLE 8.2
EPA Standard Default Exposure Factors

Land Use	Exposure Pathway	Daily Intake	Exposure Frequency (days year^{-1})	Exposure Duration (years)
Residential	Ingestion of potable water	2 L day^{-1}	350	30
	Ingestion of soil and dust	200 mg (child)	350	6
		100 mg (adult)		24
	Inhalation of contaminants	20 m^3 (total)	350	30
		15 m^3 (indoor)		
Industrial and commercial	Ingestion of potable water	1 L	250	25
	Ingestion of soil and dust	50 mg	250	25
	Inhalation of contaminants	20 m^3 (workday)	250	25
Agricultural	Consumption of homegrown produce	42 g (fruit) 80 g (vegetable)	350	30
Recreational	Consumption of locally caught fish	54 g	350	30

Source: Reprinted with permission from Ian Pepper, Charles Gerba and Mark Brusseau, eds., *Pollution Science* (New York: Academic Press, 1996), 350.

scenarios, e.g., in soil, air, and food—let us assume that monitoring wells have been installed at six locations on the Superfund site. The arsenic concentrations in groundwater samples might be as follows: Monitoring well (MW) #1: 1,367 micrograms per liter or parts per billion (ppb); MW #2: 241 ppb; MW #3: 15 ppb; MW #4: 522 ppb; MW #5: 945 ppb; and MW #6: 59 ppb. (Note: The federally designated maximum contaminant level [MCL] for arsenic in drinking water is 10 ppb.) Given the range of arsenic concentrations in the six monitoring wells, arriving at a single number to represent arsenic concentration is as much a matter of policy as it is of science.

A conservative policy approach would be to use the highest arsenic concentration among the well samples, 1,367 ppb. Another conservative approach would be to determine the arithmetic mean and standard deviation of arsenic concentrations from all the monitoring wells and use the upper 95% confidence limit, i.e., a value equal to the mean plus two standard deviations. The arithmetic mean of the six monitoring wells is 525 ppb, the standard deviation is 537 ppb, and the upper 95% confidence limit is 1,599 ppb, which is greater than the highest concentration. Less conservative policy approaches might use the arithmetic mean, 525 ppb, or the geometric mean, 229 ppb. Depending on the particular law as well as agency regulations and policies governing a particular risk assessment, the exposure to arsenic in groundwater, and hence the risk to human health, might vary by up to a factor of 7 in this hypothetical example. The exposure estimate could be refined somewhat by installing groundwater monitoring wells at additional locations around the site. However, each monitoring well costs thousands of dollars, and the odds of achieving a significant improvement in the exposure estimate must be weighed against the cost.

Let us calculate the chronic daily intake of arsenic by a future resident on the Superfund site if the groundwater aquifer were not cleaned up. The CDI is calculated per kilogram of body weight and then averaged over a person's lifetime:

$$CDI = \frac{(concentration) \times (daily\ intake) \times (duration) \times (frequency) \times (bioavailability)}{(body\ weight) \times (averaging\ time)}$$

where
concentration = concentration of arsenic in groundwater; if the upper 95% confidence limit is greater than the highest detected concentration, then the highest concentration might be used, in this example 1,367 ppb or 1.367 mg/L
daily intake = 2 L/day (EPA default value in Table 8.2)
duration = duration of exposure; a period of 30 years is assumed
frequency = frequency of exposure; an ingestion frequency of 350 days per year is assumed
bioavailability = fraction of arsenic that is taken up into the bloodstream; a conservative bioavailability fraction of 1.0 (100%) is assumed
body weight = 70 kg, which is the weight of an average adult male
averaging time = lifetime of 70 years or 25,550 days

Human Health Risk Assessment

then

$$CDI = \frac{(1.367 \text{ mg/L}) \times (2 \text{ L/day}) \times (30 \text{ years}) \times (350 \text{ days/year}) \times (1.0)}{(70 \text{ kg}) \times (25{,}550 \text{ days})}$$

$$= 0.0161 \text{ mg/kg/day}$$

The chronic daily intake is used to predict both the risk of noncancer health effects and the risk of cancer. Analogous estimates of chronic daily intake can be made for other pathways of exposure in other risk scenarios. For example, exposure to an air pollutant could be estimated as follows:

$$CDI = \frac{(\text{concentration}) \times (\text{inhalation rate}) \times (\text{duration}) \times (\text{frequency}) \times (\text{bioavailability})}{(\text{body weight}) \times (\text{averaging time})}$$

where
  concentration = concentration of toxic chemical in the air (examples: benzene, particulate matter, ozone) in units of milligrams per cubic meter (mg/m^3)
  inhalation rate = 20 m^3 (EPA default value, see Table 8.2)
  duration = years at-risk population is assumed to be exposed, e.g., 30 years
  frequency = frequency at-risk population is assumed to be exposed, e.g., 350 days per year
  bioavailability = fraction of inhaled toxic chemical that is taken up into the bloodstream, e.g., 1.0 (100%) if there are no studies documenting that bioavailability is lower
  body weight = average weight of exposed population, e.g., 70 kg for adult males, less for children
  averaging time = approximate lifespan of members of exposed population in days, e.g., 25,550 days if lifespan is assumed to be 70 years

Other exposure pathways include the ingestion of contaminated food, skin contact with a hazardous chemical, and the ingestion of contaminated dirt. The CDI for each of these pathways is estimated analogously to the CDI for ingesting contaminated water or breathing contaminated air. However, small children are assumed to ingest more soil than adults, e.g., 200 mg/day as opposed to 100 mg/day (Table 8.2), and of course children weigh less on average, about 20 kg instead of 70 kg. Depending on the specific assumptions of increased soil ingestion and lower body weight used in the exposure analysis, children's estimated CDI can be an order of magnitude (factor of 10) greater than adults' CDI by this pathway.

8.4.2 Biomonitoring

A second approach to exposure analysis is biomonitoring, in which members of a population suspected to have been exposed to a toxic chemical donate tissue samples such as blood, urine, or hair, and the samples are analyzed in a state-certified testing laboratory. The greatest amount of exposure information can generally be obtained from blood samples. In recent years, biomonitoring studies in the United States and the European Union have documented the presence, in the blood of children and adults, of dozens of chemicals that are known to be persistent in the environment, bioaccumulative, and toxic (called PBT) (Chapters 4 and 10). Although biomonitoring has an important role in raising awareness of low-dose, long-term exposure to industrial chemicals and their potential for adverse health effects, blood levels of chemicals are rarely used in quantitative risk assessments. The reason is that risk-assessment calculations are based on animal toxicity studies, where the administered, or external, dose of a toxic chemical is known but the internal dose (the blood level of the toxic chemical) generally is not.

More research is needed to understand the long-term implications of the body burdens of multiple toxic chemicals that biomonitoring studies have revealed are present in every person in the world and in many species of wildlife. Better understanding of the relationship between chemical body burdens and disease would make it possible to use biomonitoring data to assess risk. It is up to national governments to decide if and how the qualitative information provided by biomonitoring studies is to be incorporated into risk-management policies (Chapter 10).

8.5 ANALYSIS OF EFFECTS

The U.S. Environmental Protection Agency (EPA) divides the health effects of toxic chemicals into two broad categories for risk-assessment purposes: risk of noncancer (noncarcinogenic) health effects and risk of cancer (carcinogenic risk) (Chapter 7). The same analysis of exposure is used for both noncarcinogenic and carcinogenic risk; however, the relationship of exposure to effect is analyzed differently for noncancer and carcinogenic risks.

8.5.1 Noncancer Health Effects

The analysis of noncancer health effects assumes that there is a threshold level of exposure beyond which members of an at-risk population will experience disease. Risk is assessed by comparing the disease threshold to chronic daily intake (CDI). If the CDI exceeds the threshold, the exposed population is considered to be at risk for noncancer health effects. The degree of risk is not quantitated. Rather, any CDI value greater than the exposure threshold can trigger regulatory action designed to manage the risk. There is some regulatory "wiggle room" to act in cases where exposure is just below the disease threshold or not to act if exposure is marginally above it. In borderline cases, the decision to pursue regulatory action is based on the overall weight of evidence, including qualitative evidence of risk that is reviewed in the final step of the risk-assessment process, the risk characterization.

How is the exposure threshold determined for a noncancer health effect? The exposure threshold is estimated on the basis of toxicity testing in animals, specifically the "no observed adverse effect level" (NOAEL) on the dose–effect curve (Point E in Figure 5.2). The NOAEL in rats or mice is not, itself, used as the threshold dose for toxic effects in humans. Rather, the NOAEL is divided by at least two

Human Health Risk Assessment

and sometimes three factors of 10 to arrive at an estimate of the human threshold for disease. These factors are known as uncertainty factors or safety factors. One factor of 10 accounts for the uncertainty involved in extrapolating from animal studies to humans. A second factor of 10 accounts for the human-to-human variation in susceptibility to chemical disease, i.e., some people are more vulnerable to the effects of toxic chemicals than others (see Figure 3.1). The number obtained by dividing the NOAEL by uncertainty factors is called the reference dose (RfD):

$$\text{RfD} = \text{NOAEL}/[10 \text{ (animal-human variation)} \times 10 \text{ (human-human variation)}] \quad (8.1)$$

A third uncertainty factor of 10 is added if the animal toxicity data do not define the NOAEL conclusively. For example, animal toxicity data may define a LOAEL (lowest observed adverse effect level) but not a NOAEL. In such a case, the reference dose might be estimated from the LOAEL by including an additional uncertainty factor of 10:

$$\text{RfD} = \text{LOAEL}/[10 \text{ (animal-human variation)} \times 10 \text{ (human-human variation)} \\ \times 10 \text{ (animal data uncertainty)}] \quad (8.2)$$

The NOAEL has been criticized because it places excessive weight on a single point on the dose-response curve. An alternative approach to estimating the reference dose is called the benchmark dose (BMD). A low incidence of toxicity is chosen, for example, manifestation of the specified toxic effect in 5% of the test population, and the upper and lower 95% confidence limits are determined by statistical modeling. The upper 95% confidence limit corresponds to a lower dose triggering the toxic effect in 5% of the population, while the lower 95% confidence limit corresponds to a higher dose causing the same incidence of morbidity. The benchmark dose corresponds to the upper 95% confidence limit, i.e., the more conservative estimate of risk. Similarly to the NOAEL, the benchmark dose, is divided by multiple uncertainty factors to estimate the reference dose:

$$\text{RfD} = \text{BMD}_x/[10 \text{ (animal-human variation)} \times 10 \text{ (human-human variation)} \\ \times 10 \text{ (data uncertainty)}]$$

where BMD_x is the benchmark dose for a specified percent response, e.g., 1%, 5%, 10%.

8.5.1.1 The Two Approaches

The NOAEL and the benchmark dose generally produce similar estimates of the reference dose. Each approach offers certain advantages, and the choice of approach depends on the strengths and weaknesses of the animal toxicity data set on which it is based. Human epidemiological data, when available, may complement but not replace the NOAEL or benchmark dose in estimating a chemical's reference dose.

The reference dose for a toxic chemical—an equivalent term is the acceptable daily intake (ADI)—is taken to represent the threshold level of exposure above which a noncancer health effect can result in humans. The hazard index (HI) indicates whether the exposure threshold is exceeded and is equal to the ratio of the chronic daily intake to the reference dose:

$$HI = CDI/RfD \qquad (8.3)$$

If the hazard index is greater than 1, i.e., if exposure exceeds the estimated threshold for a specified noncancer health effect, then the exposed population is considered to be at risk. If a risk scenario includes several pathways of exposure to a chemical of concern, a hazard index is calculated separately for each pathway, and the results are summed to find the total hazard index for that chemical.

8.5.1.2 Air Quality Index

Airborne toxic chemicals may pose a risk of toxic effects directly on the lungs as well as toxic effects on other organs and tissues after entering the bloodstream through the lungs. The reference dose for an air pollutant is termed the reference concentration, or RfC. The reference concentration differs from the reference dose in that it is the actual concentration of a chemical in the air, not the average daily dose of chemical that enters the bloodstream. The air quality index (AQI) is an estimate of the risk of effects from air pollutants, particularly in people with pre-existing conditions such as asthma. The air quality index is calculated as the ratio of the concentration of an air pollutant to its reference concentration multiplied by 100:

$$AQI = [(\text{chemical concentration in the air})/RfC] \times 100$$

The concentrations of five major air pollutants are regulated under the Clean Air Act: ground-level ozone, particulate matter (PM), carbon monoxide, sulfur dioxide, and nitrogen dioxide. The concentrations of these five pollutants are monitored continuously through a national network of monitoring stations maintained by the EPA and the National Oceanic and Atmospheric Administration (NOAA). The air quality index is tracked for each pollutant using six categories to characterize air quality and risks to human health. An air quality index value of 0 to 50 corresponds to good air quality; 51 to 100, moderate air quality; 101 to 150, unhealthy for sensitive groups; 151 to 200, unhealthy; 201 to 300, very unhealthy; and 301 to 500, hazardous. The Clean Air Act mandates that in large cities with populations greater than 350,000 inhabitants, state and local agencies provide the public with daily reports of the air quality index. The highest AQI value of the five regulated pollutants is reported as the AQI value for the day.

To summarize the noncancer health risk-assessment process, chemicals of concern, pathways of exposure, and exposed populations are identified in the first step of the risk assessment, hazard identification. In the second step, analysis of exposure, the doses are estimated for each population, each exposure pathway, and each chemical of concern in the form of chronic daily intakes, or CDIs. In the third step of the risk assessment, analysis of effect, noncancer health effects are estimated by comparing CDIs to reference doses, or RfDs, derived from animal toxicity studies (with input from human epidemiological studies, when available). If the CDI is greater

Human Health Risk Assessment

than the RfD, the exposed population is considered to be at risk for noncancer health effects.

8.5.1.3 Other Strategies for Noncancer Risk Assessment

Two other approaches are used to estimate noncancer risks from toxic chemicals: margin of exposure and therapeutic index. Both are less formal and more approximate than the hazard index. The margin of exposure (MOE) is the ratio of the NOAEL in an animal toxicity study to the CDI projected for a human population. Uncertainty factors are omitted from the calculation. For example, if the NOAEL for reproductive toxicity were, say, 1.5 mg/kg/day, and the CDI were 0.003 mg/kg/day, then the MOE would be 500. The MOE is interpreted by comparing it to a margin of safety (MOS) established by a government agency. In general, an MOE of less than 100 is considered to be cause for concern.

The therapeutic index (TI) is used to evaluate the safety of drugs. The therapeutic index is calculated by dividing the median toxic dose of a drug (TD_{50}) by its median effective dose (ED_{50}) (see Figures 3.1 and 5.1 and Chapters 3 and 5 for an explanation of median dose). A therapeutic index of 10 is generally considered to represent an acceptable margin of safety, i.e., toxic doses of a drug are approximately 10 times higher than therapeutic doses. A number of common drugs have a therapeutic index less than 10, and patients taking these drugs must be monitored carefully to avoid adverse ("side") effects.

8.5.2 CANCER RISK

The assessment of noncancer health risks is based on the concept that low levels of exposure are acceptable, and that populations manifest health effects only after exposure thresholds are crossed. Determining exposure thresholds involves a significant amount of scientific uncertainty; however, a large body of evidence supports the hypothesis that thresholds do, in fact, exist for most if not all noncancer health effects. In the case of cancers, on the other hand, a good deal of evidence points to the absence of exposure thresholds. While conclusive proof is lacking, it is plausible, based on what is known about the genetic origins of cancer, that a single molecule of a toxic chemical could cause a mutation in DNA that triggers the process of carcinogenesis (Chapter 7). In other words, if it were possible to perform an animal toxicity study at ultra-low doses of a chemical carcinogen (note that it's not really possible because it would require thousands if not tens of thousands of animals and be far too expensive; see Chapter 5), the dose–effect curve that would be obtained at low levels of exposure can plausibly be predicted to pass through zero. While the no-threshold hypothesis is plausible, other models of carcinogenesis predict that more than one "hit" on a DNA molecule is required to start the cancer process. The issue is far from settled, and the origins of cancer remain an active field of scientific research.

The EPA bases carcinogenic risk assessment on the conservative hypothesis that there is no exposure threshold for cancer. This no-threshold assumption is combined with a second hypothesis: that cells develop cancer in a series of steps or stages resulting from how they change in response to the effects of various growth factors, e.g., proto-oncogenes and hormones (Chapter 7). These two assumptions provide the basis for EPA's preferred model for cancer risk assessment, the linearized multistage model of carcinogenesis. The statistical methods used to derive the linearized multistage model are beyond the scope of this book. Essentially, the model is used to extrapolate the dose–effect curve from the high doses in animal toxicity studies to the low-dose range typical of human exposures. Human epidemiological studies, if available, may also be used as a starting point for extrapolation. This means that the dose–effect curve constructed from the results of an animal toxicity study or a human epidemiological study is extrapolated into a region of low-dose exposure that is hundreds or thousands of times lower than the doses used in the study and where quite literally no data exist.

Needless to say, there is considerable uncertainty as to the shape and slope of the extrapolated curve, particularly when animal toxicity data are used, which is generally the case. The uncertainty is expressed by bounding the extrapolated curve with 95% upper and lower confidence limits. The upper 95% confidence limit corresponds to a higher incidence of cancer at a given dose, while the lower 95% confidence limit corresponds to a lower incidence. The confidence limits parallel the extrapolated curve and effectively show the range of cancer risk. The range of the estimated risk, i.e., how far away the confidence limits are from the extrapolated curve, depends on the quality and quantity of the high-dose data. To provide added protection for public health, the EPA uses the upper 95% confidence limit to calculate carcinogenic risk. The upper 95% confidence limit turns out to be a straight line in the statistical model used by the EPA. In other words, the linearized multistage model says that cancer risk, estimated conservatively, increases as a linear function of dose at low levels of exposure. The slope of the extrapolated straight line indicates how much cancer risk increases with exposure. The steeper the slope, the more carcinogenic potency a chemical has and the more risk increases with each incremental increase in exposure.

How is the linearized multistage model of carcinogenesis applied to the assessment of cancer risk? The straight line that results from extrapolating the upper 95% confidence limit to the low-dose range (where there are no data) has the following simple mathematical form that may be familiar from high school algebra:

$$y = mx + b$$

where x is the value of the independent variable on the x-axis, y is the value of the dependent variable on the y-axis, m is the slope, and b is the point of intersection of the straight line with the y-axis. In the context of carcinogenic risk assessment, x is the dose of the carcinogenic chemical to which an at-risk population is exposed, y is the incidence of cancer in the at-risk population above the background incidence, m is the rate of increase in cancer incidence as a function of dose, and b is equal to zero (in the linearized multistage model, the dose–effect curve passes through zero). Since $b = 0$, the slope reduces to:

$$m = y/x$$

Substituting terms used in cancer risk assessment:

Human Health Risk Assessment

$$q_1^* \text{ (mg/kg/day)}^{-1} = \text{Cancer risk/CDI (mg/kg/day)} \quad (8.4)$$

$$\text{Cancer risk} = q_1^* \text{ (mg/kg/day)}^{-1} \times \text{CDI (mg/kg/day)} \quad (8.5)$$

where q_1^* is the slope of the straight line corresponding to the upper 95% confidence limit at low doses and is called the slope factor or cancer potency factor, CDI is the chronic daily intake averaged over a lifetime, and cancer risk is equivalent to the incidence of cancer above background. In other words, risk is expressed as the fraction of the population that is predicted to contract cancer as a result of low-dose exposure to the chemical of concern, for example, 5 per thousand (one two-hundredth of the population, or 0.5%) or 2 per million (one five hundred thousandth of the population, or 2×10^{-6}). Because incidence is a fraction or percent, it does not have units. The units of the slope factor are the reciprocal of the units of CDI, $(\text{mg/kg/day})^{-1}$.

Slope factors can be used to compare the relative cancer potencies of toxic chemicals. That is because when the CDI is arbitrarily set equal to 1 mg/kg/day, cancer incidence is equal to the slope factor (Equation 8.5). Another way to say this is that the slope factor gives the cancer incidence per unit (mg/kg/day) of exposure. This level of cancer risk, i.e., the risk from exposure to 1 mg/kg/day of a carcinogenic chemical averaged over a lifetime, is called the unit risk. Unit risks are numerically equal to slope factors. They provide a convenient way to compare the estimated carcinogenic potencies of different chemicals.

8.5.3 Risk Calculations

The chronic daily intake (CDI) estimated in the analysis of exposure, the second step of the risk assessment, is used to calculate the risks of both noncancer health effects and cancer. Risk calculations are also referred to as "quantitative risk assessment," a term that is somewhat misleading because the word "quantitative" implies a high degree of accuracy, which is clearly not the case. In the first risk scenario described in Section 8.3, future residents drink arsenic-contaminated water from the aquifer beneath a former Superfund site. Their CDI by this pathway is estimated to be 0.0161 mg/kg/day of arsenic. The oral reference dose (RfD) for arsenic is 3×10^{-4} mg/kg/day, according to the EPA's Integrated Risk Information System (IRIS) (U.S. EPA 2009). The hazard index (HI) for noncancer health effects caused by this chemical of concern by this exposure pathway is calculated using Equation (8.3):

$$\begin{aligned}
\text{HI} &= \text{CDI/RfD} \\
&= 1.61 \times 10^{-2}/3 \times 10^{-4} \\
&= 0.54 \times 10^2 \\
&= 54
\end{aligned}$$

The hazard index greatly exceeds a value of 1, indicating substantial risk of noncancer health effects if future residents drank water from the aquifer underneath the Superfund site.

The cancer risk from ingesting 0.0161 mg/kg/day of arsenic in drinking water is calculated using Equation (8.5). The slope factor, or cancer potency factor, for arsenic is 1.5 (mg/kg/day)$^{-1}$ (IRIS). Therefore, the cancer risk is calculated to be:

Cancer risk = Slope factor × CDI
= 1.5 (mg/kg/day)$^{-1}$ × 0.0161 mg/kg/day
= 0.024
= 2.4×10^{-2}

The incidence of cancer above background is estimated to be 2.4% of the exposed population (24 out of every thousand people, or 24,000 out of every million people). This is a very high level of risk. EPA generally regulates carcinogenic chemicals to a cancer risk level of one in ten thousand to one in a million (10^{-4} to 10^{-6}).

Risks from other pathways of exposure and/or other chemicals of concern are considered to be additive unless there is evidence that the toxicities of two or more chemicals are synergistic (i.e., enhance each other so that risk is greater than the sum of the risk from either chemical alone) or inhibitory (i.e., interfere with each other so that risk is less than the sum of the risk from either chemical alone). Very little is known about the interactions between toxic chemicals, and risks from multiple chemicals and multiple exposure pathways are usually added together to obtain an estimate of total risk. In the case of noncancer health risk, the hazard index (HI) is calculated separately for each chemical and each exposure pathway, and total risk is equal to the sum of the HI values from all chemicals and all pathways. In the case of cancer risk, the cancer incidence is calculated for each chemical and each exposure pathway, and total risk is equal to the sum of the cancer incidences from all chemicals and all pathways. Cancer risk is the probability of getting cancer (morbidity), not the probability of dying from cancer (mortality). Many people get cancer and survive.

8.6 RISK CHARACTERIZATION

The final step of the risk-assessment process, risk characterization, is a narrative that incorporates all the information assembled in the previous three steps. It also marshals qualitative evidence of risk that is not included in the formal risk-assessment calculations. The narrative weighs all the evidence and uses professional judgment to draw conclusions regarding risk.

8.6.1 UNCERTAINTY

Discussing the uncertainties involved in risk calculations is an important aspect of risk characterization. "Quantitative risk assessments" are fraught with uncertainty. Estimates of exposure (chronic daily intake) are probably accurate to roughly a factor of 10. Toxicity values—the reference dose for noncancer health effects and the slope factor for cancer risk—are also essentially order-of-magnitude estimates. As a result of these uncertainties, "quantitative risk assessment" gives a number that could be reasonably accurate but that could also underestimate or overestimate risk by a factor of 10 to 100 or more. Uncertainty of this magnitude is problematic; however, it does not follow that risk-assessment calculations are worthless. They are "ballpark" estimates that provide a rational basis for prioritizing risks and allocating limited resources to manage them. From the standpoint of managing risks rationally, it is far better to have ballpark estimates than to have no risk estimates at all. Factoring the high degree of uncertainty into risk-management decisions and the communication of risk to the general public is an ongoing challenge for risk professionals.

8.6.2 Weight of Evidence

Qualitative evidence may be available that is relevant to assessing risk but cannot be included in a quantitative calculation. Qualitative evidence of risk is marshaled separately and considered alongside the results of risk calculations. Examples include fate and transport data suggesting exposure pathways in addition to those included in the risk calculation; evidence for toxic effects besides those included in the risk calculations; the strength of evidence indicating that a chemical included in the carcinogenic risk assessment is, in fact, carcinogenic in humans; evidence regarding additive, synergistic, or inhibitory interactions among chemicals of concern; and gaps in toxicity data.

Risk assessors apply professional judgment to weigh the quantitative and qualitative evidence of risk. Their conclusions, narrated in the risk characterization step, provide the basis for managing toxic chemical risk within the framework of applicable laws and agency regulations and policies.

8.6.3 Limited Information on Chemical Toxicities

The number of chemicals that have been characterized in animal toxicity studies or human epidemiological studies is tiny compared with the approximately 80,000 chemical compounds in the TSCA Inventory, i.e., reported as being in commerce in the United States. A search of the EPA's Integrated Risk Information System (IRIS) (U.S. EPA 2009) yields 287 chemicals that have been characterized with respect to carcinogenicity. Of these, 97 chemicals, or 34%, are classified as known or probable/likely to be carcinogenic in humans. It seems unlikely that 34% of all chemicals in commerce are carcinogenic, because chemicals that are studied tend to be those that have reason to be suspected of toxicity. Nevertheless, if only 1% of the 80,000 chemicals in commerce were carcinogenic, that would add up to 800 carcinogens, or some 700 more than the 97 that are currently known. The point is that a general lack of toxicity data, not just on carcinogenicity, but on noncancer health impacts, including reproductive, developmental, and neurological effects, undermines the risk-assessment process. Data gaps are the single biggest obstacle to performing useful risk assessments and will remain so until more resources are invested in characterizing the toxicity of everyday chemicals.

STUDY QUESTIONS

1. Name the four steps of the human health risk-assessment process and briefly outline each step.
2. What is the chronic daily intake (CDI)?
3. Explain the major sources of uncertainty in estimating the CDI.
4. What is the basis for assessing carcinogenic risk?
5. What is the slope factor and how is it derived?
6. Describe three sources of uncertainty in carcinogenic risk assessment.
7. What is the unit risk?
8. What is the reference dose (RfD) and how is it derived? What is the acceptable daily intake (ADI)? What is the reference concentration (RfC)?
9. What is the hazard index (HI)? How is the hazard index used to assess risk?
10. How is the hazard index concept applied to the management of health risks from poor air quality?
11. Describe three sources of uncertainty in assessing noncancer health effects.
12. Name the greatest single source of uncertainty in human health risk assessments.
13. Define the margin of exposure (MOE) and the therapeutic index. How are they used to assess risk?
14. Methylmercury damages the nervous system, and the developing nervous system of embryos and young children up to the age of seven years is particularly vulnerable. Mercury is present in coal and is emitted to the air from coal-burning power plants. It is transported hundreds or thousands of miles by air currents and falls on oceans and lakes, where it enters aquatic food webs. According to data gathered by the U.S. Food and Drug Administration from (FDA) 2002–2004, the mean concentration of mercury in 228 samples from all species of tuna fish was 0.383 parts per million (ppm), the standard deviation was 0.269 ppm, and the maximum concentration was 1.300 ppm (http://www.cfsan.fda.gov/~frf/sea-mehg.html). The FDA and EPA advise women of child-bearing age that it is safe to consume up to two meals (12 oz.) of canned light tuna per week but only one meal (6 oz.) of albacore ("white") tuna per week (http://www.cfsan.fda.gov/~dms/admehg3.html). Assuming that a young woman weighing 57 kg eats two meals of tuna per week but does not distinguish among different species of tuna, that the mercury level in the tuna approximates the mean concentration in the FDA study, and that all of the mercury in the fish is methylmercury, calculate the hazard index for noncancer health effects. The RfD for methylmercury is 0.0001 mg/kg/day (http://www.epa.gov/iris/subst/0073.htm). What is the hazard index if it is assumed that the fish contain the maximum concentration of mercury measured in the FDA study? What is the carcinogenic risk if maximum exposure to mercury is assumed?
15. A horizontal gas well is drilled in the Marcellus Shale in south-central New York State and stimulated by injecting 5 million gallons of hydraulic

fracturing fluid ("fracking fluid") under high pressure to fracture the shale rock and release the gas. Some of the fracking fluid, which contains a number of organic chemicals, including benzene, finds its way into a nearby aquifer and contaminates a homeowner's private drinking water well. Assuming the concentration of benzene in the aquifer averages 4.6 parts per billion (ppb), what is the risk of the homeowner contracting leukemia, the type of cancer caused by benzene? What is the risk of a decrease in the number of lymphocytes (a type of immune cell) in the blood, the most prominent noncancer health effect of benzene?

ANSWERS TO STUDY QUESTIONS

1. The four steps of the risk-assessment process are hazard identification, analysis of exposure, analysis of effect, and risk characterization. In the hazard identification step, the risk assessor identifies chemicals of concern, environmental pathways of exposure, and populations and subpopulations at risk. The exposure analysis develops exposure scenarios and estimates the chronic daily intake of each chemical of concern. In the analysis of effect, the risk assessor combines the chronic daily intake calculated in the exposure analysis with toxicity data from animal studies (and/or human epidemiological studies, if available) to estimate the risk of toxic effects in exposed populations, whereby risks to public health are divided into two broad categories: noncancer health effects and cancer. The final step of the risk-assessment process, risk characterization, is a narrative that marshals all the evidence of risk to public health, including quantitative risk assessments and qualitative evidence of risk. The risk assessor weighs all the evidence and uses professional judgment to draw conclusions about risks.
2. The chronic daily intake (CDI) is the amount, or dose, of a chemical that reaches the bloodstream each day per kilogram of body weight averaged over the total number of days a person is exposed. CDI is a product of several factors: the concentration of chemical in the food or environmental medium (water, air, dirt) with which a person comes in contact; the degree of contact, e.g., how much water a person drinks or how much air she breathes; and bioavailability, i.e., the fraction of chemical in the environmental medium or food that enters the bloodstream. CDI is also affected by the body weight that is assumed for members of the at-risk population. See Section 8.4.1 for equations defining CDI for exposure by ingestion and inhalation.
3. The concentration of a chemical of concern typically varies widely among environmental samples. Further, the degree of contact with a receptor, i.e., a population of organisms or an ecosystem, fluctuates over time. Finally, accurate bioavailability data are not available for many chemicals and routes of exposure. Thus, exposure analysis involves considerable uncertainty, and small differences in assumptions can result in significant differences in estimated exposure and hence significant differences in estimated risk.

4. Carcinogenic risk is assessed on the basis of the incidence of cancer above background in a population exposed to a chemical of concern. The incidence of morbidity (disease), not mortality (death), is used to assess the risk of cancer.
5. The slope factor, also called the cancer potency factor, corresponds to the rate of increase in cancer incidence as a function of increasing exposure. The greater the slope factor, the greater is the number of additional cancer cases with each increase in exposure, and therefore the more potent the chemical is as a carcinogen. The slope factor is derived by extrapolating the dose–effect curve obtained in a high-dose carcinogenicity study in animals, typically rats or mice, to the low-dose range characteristic of human exposure. Extrapolation is based on the EPA's linearized multistage model of carcinogenesis, which assumes that there is no exposure threshold for cancer, and therefore the dose–effect curve is extrapolated through the origin of the x–y coordinate system (i.e., through zero). The slope factor corresponds to the slope of the upper 95% confidence limit of the extrapolated curve, which is a straight line in the EPA model.
6. One source of uncertainty is whether a chemical causes cancer in humans. In particular, a chemical may cause cancer in animals but not in humans. Another source of uncertainty is the extrapolation of the dose–effect curve from high doses in animal toxicity studies to low doses, where generally no data exist and where the relationship between dose and cancer incidence is based almost entirely on conceptual models of carcinogenesis. A third source of uncertainty is the possible effects of interactions among multiple chemicals to which a population may be exposed.
7. The unit risk is the increase in cancer incidence that results from an increase in the chronic daily intake (CDI) of 1 mg/kg/day of a carcinogenic chemical in food or soil. For drinking water, the unit risk is defined as the increase in cancer incidence resulting from a CDI increase of 1 μg/kg/day (1 μg is equal to one microgram, or one thousandth of a milligram). The unit risk is equal to the slope factor (see Equation 8.5).
8. The reference dose (RfD) assumes there is a threshold of exposure below which a chemical does not produce a toxic effect because the body is able to detoxify and/or eliminate it. The reference dose is derived either from a "no observed adverse effect level" (NOAEL) or from a benchmark dose (BMD) determined in an animal toxicity study. The NOAEL or BMD is divided by at least two uncertainty factors or safety factors: a factor of 10 to account for the uncertainty involved in extrapolating from animals to humans, and a second factor of 10 to account for variation in human sensitivity. If the animal toxicity data supporting the NOAEL or BMD are not definitive, a third safety factor of 10 is included. Thus, the RfD is set equal to the NOAEL or BMD divided by 100; alternatively, it is set equal to a number approximating the NOAEL or BMD divided by 1,000. The acceptable daily intake (ADI) is the same as the reference dose. The reference concentration (RfC) refers to the concentration of a pollutant in the air. It differs from the

reference dose in that it refers to the concentration of a chemical of concern rather than to the dose that is taken up into the bloodstream.

9. The hazard index (HI) is used to assess the risk of a noncancer health effect in an exposed population. The hazard index is calculated by dividing the chronic daily intake (CDI) by the reference dose (RfD): HI = CDI/RfD. A hazard index equal to or greater than 1 indicates significant risk. The higher the hazard index, the greater is the risk.

10. The air quality index is analogous to the hazard index and is used to assess risks from adverse health effects caused by short-term exposure to air pollutants. The concentrations of five major air pollutants regulated under the Clean Air Act are tracked daily by a nationwide network of monitoring stations located in population centers with more than 350,000 people. The air quality index is calculated daily for each pollutant by dividing its concentration in the air by its reference concentration (RfC) and multiplying the result by 100. An air quality index of 100 or more triggers public health warnings.

11. The greatest single source of uncertainty in the analysis of effect is the complete lack of animal toxicity data for over 90% of chemicals in commerce. For the minority of chemicals for which animal toxicity studies have been performed, the major sources of uncertainty are the applicability of animal data to humans, the variation in sensitivity within the human population, and the quality of the animal data.

12. The single greatest source of uncertainty in human health risk assessments is the lack of animal toxicity data on most chemicals with regard to noncancer health effects as well as carcinogenicity. Over 90% of chemicals in commerce lack the toxicological data required to perform meaningful risk assessments.

13. The margin of exposure and the therapeutic index are less formal and more approximate estimates of risk than the hazard index or the unit risk. The margin of exposure is equal to the ratio of the exposure that produces a toxic effect in an animal study to the exposure estimated for a human population. A margin of exposure of less than 100 is generally considered cause for concern. The therapeutic index is the ratio of the dose of a therapeutic drug that produces a toxic effect to the dose that produces a therapeutic effect. Drug doses are defined as median doses. Thus, the median effective dose (ED_{50}) is the dose that produces the therapeutic effect in 50% of the population. The median toxic dose (TD_{50}) is the dose that produces the toxic effect in 50% of the population. The therapeutic index is calculated as follows: TI = TD_{50}/ED_{50}. The therapeutic index provides a general indication of a drug's margin of safety and is useful in protecting patients from the adverse ("side") effects of therapeutic drugs.

14. Calculate the chronic daily intake and divide it by the reference dose to obtain the hazard index:

$$\text{CDI} = \frac{(\text{concentration}) \times (\text{daily intake}) \times (\text{duration}) \times (\text{frequency}) \times (\text{bioavailability})}{(\text{body weight}) \times (\text{averaging time})}$$

Concentration = mean methylmercury concentration for all tuna in FDA study
= 0.383 ppm
= 0.383 mg/kg
Daily intake of tuna = 12 oz./week × 29.53 g/oz. × 1 week/7 days × 1 kg/1,000 g
= 0.051 kg/day
Duration of exposure = 30 years (EPA default value; see Table 8.2)
Frequency of exposure = 2 days/week × 52 weeks/year = 104 days/year
Bioavailability = 0.95 (http://www.epa.gov/ncea/iris/subst/0073.htm)
Body weight = 57 kg
Averaging time = 70 years

$$\text{CDI} = \frac{(0.383 \text{ mg/kg}) \times (0.051 \text{ kg/day}) \times (30 \text{ years}) \times (104 \text{ days/year}) \times (0.95)}{(57 \text{ kg}) \times (70 \text{ years} \times 365 \text{ days/year})}$$

= 0.00004 mg/kg/day

The chronic daily intake is 0.00004 mg/kg/day or 0.04 μg/kg/day of methylmercury. The reference dose for methylmercury in women of childbearing age is 0.0001 mg/kg/day or 0.1 μg/kg/day. Therefore, the hazard index is:
HI = CDI/RfD = (0.04 μg/kg/day)/0.1 μg/kg/day
HI = 0.4

If exposure to the maximum concentration of methylmercury found in the FDA study, 1.300 ppm, is assumed, then the chronic daily intake is increased by the ratio of the maximum concentration to the average concentration: (1.300 mg/kg)/(0.383 mg/kg) = 3.39. The hazard index is increased by the same factor, to (3.39) × (0.4) = 1.4. A hazard index value greater than 1.0 is cause for concern and is consistent with the joint FDA/EPA advisory to women of childbearing age not to consume more than one meal of albacore ("white") tuna per week, since albacore tuna tends to have high levels of mercury.

Methylmercury is classified as a possible human carcinogen. This means that there is little evidence to suggest that methylmercury causes cancer in humans. A slope factor has not been determined.

15. The cancer risk can be estimated directly from the unit risk. The unit risk for benzene in drinking water is 4.4×10^{-7} to 1.6×10^{-6} per μg/L according to IRIS (http://www.epa.gov/ncea/iris/subst/0276.htm#quaoral). If the benzene concentration is 4.6 μg/L (ppb), the cancer risk is 2×10^{-6} to 7.4×10^{-6}. Stated differently, the cancer incidence above background is predicted to be

between two and seven cases per million people exposed. The cancer risk could also be estimated by calculating the chronic daily intake and multiplying it by the oral slope factor, which is 1.5×10^{-2} to 5.5×10^{-2} (IRIS).

To estimate noncancer health risk, calculation of the chronic daily intake cannot be avoided:

$$CDI = \frac{(\text{concentration}) \times (\text{daily intake}) \times (\text{duration}) \times (\text{frequency}) \times (\text{bioavailability})}{(\text{body weight}) \times (\text{averaging time})}$$

$$CDI = \frac{(0.0046 \text{ mg/L}) \times (2 \text{ L/day}) \times (30 \text{ years}) \times (350 \text{ days/year}) \times (1.0)}{(70 \text{ kg}) \times (25{,}550 \text{ days})}$$

$$= 0.000054 \text{ mg/kg/day}$$

The oral RfD for noncancer health effects of benzene is 4.0×10^{-3} mg/kg/day.

$$HI = CDI/RfD = (5.4 \times 10^{-5})/(4 \times 10^{-3}) = 1.4 \times 10^{-2} = 0.014$$

REFERENCES

Amdur, Mary, John Doull, and Curtis Klaassen, eds. 1991. *Casarett & Doull's toxicology: The basic science of poisons*. 4th ed. New York: Pergamon Press.

Pepper, I. L., C. P. Gerba, and M. L. Brusseau, eds. 1996. *Pollution science*. New York: Academic Press.

U.S. EPA. 2009. Integrated Risk Information System (IRIS). Environmental Protection Agency. http://cfpub.epa.gov/ncea/iris/index.cfm

SUGGESTED READING

Faustman, E. M. and G. S. Omenn. 2001. Risk assessment. In *Casarett & Doull's toxicology: The basic science of poisons*, 6th ed., ed. C. D. Klaassen, 83–104. New York: McGraw-Hill.

9 Ecological Risk Assessment

9.1 FRAMEWORK FOR ECOLOGICAL RISK ASSESSMENT

Ecological risk assessments are conducted in a manner analogous to human health risk assessments: Hazards are identified, exposure and effects are analyzed, and risk is characterized. However, ecological risk assessments are generally more complex than human health risk assessments. The reader will recall that the discipline of risk assessment has borrowed the term *receptor* from biochemistry and applied it to any population of organisms or any ecosystem that is at risk from one or more chemical, physical, or biological stressor. Human health risk assessments by definition address a single class of receptors—human populations—and they consider only those stressors that directly affect human health, e.g., toxic chemicals and microbial pathogens. By contrast, ecological risk assessments may be performed on any of a host of receptors, such as animal or plant populations, communities (assemblages of populations), or whole ecosystems. In addition to chemical stressors, ecological risk assessments consider physical stressors, for example, dams that impede salmon spawning and deforestation that results in fragmentation and loss of bird habitat; they also address biological stressors such as invasive plant species that crowd out native vegetation.

The effectiveness of ecological risk assessment as a tool for managing risk depends on how well the functioning of a receptor—a population, community, or ecosystem—is understood. Human knowledge of ecological receptors and their world is generally limited. Our limitations put a premium on conceptual and procedural clarity. The U.S. Environmental Protection Agency (EPA) proposed a framework for ecological risk assessment in 1992, and this framework was finalized with the publication of the "Guidelines for Ecological Risk Assessment" (U.S. EPA 1998). Comprehensive in scope and clear and concise in presentation, the EPA guidelines have found broad acceptance and application. They describe a process characterized by: scientific rigor; extensive communication among risk assessor, risk manager, and interested parties; and fluidity in the process itself (Figure 9.1). The process employs the best available science to identify and mitigate ecological risks in the context of societal values, existing laws and regulations, and available resources inside and outside of government.

This chapter provides an overview of EPA's framework as an introduction to key concepts and approaches in assessing ecological risk. Interested readers are encouraged to consult the references at the end of this chapter for a more complete discussion of ecological risk assessment and to familiarize themselves with concrete

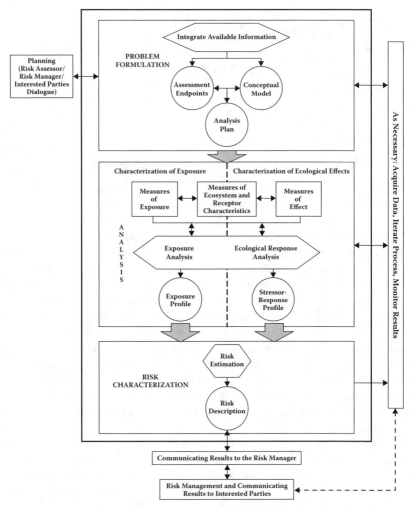

FIGURE 9.1 The U.S. EPA's framework for ecological risk assessment. Rectangles show inputs, hexagons indicate actions performed by the risk assessor, and circles name outputs produced by the risk assessor's actions. (Reprinted from U.S. Environmental Protection Agency, *Guidelines for Ecological Risk Assessment* [EPA/630/R-95/002F, http://cfpub.epa.gov/ncea/cfm/recordisplay.cfm?deid=12460, 1998], 4.)

examples of the ecological risk-assessment process as presented in the EPA's "Review of Ecological Assessment Case Studies from a Risk Assessment Perspective" (U.S. EPA 1994). Real-world case studies, which are beyond the scope of this book, offer the most effective path to understanding ecological risk assessment. This chapter offers a conceptual introduction.

9.2 THE EPA'S ECOLOGICAL RISK-ASSESSMENT PROCESS

The general process used by EPA to conduct an ecological risk assessment is diagrammed in Figure 9.1. The process is divided into four steps: planning, problem

Ecological Risk Assessment

formulation, analysis, and risk characterization. The process allows for feedback and course adjustments at all steps, as indicated by the long rectangle on the right, labeled "as necessary: acquire data, iterate process, monitor results." The EPA framework envisions a process in constant contact with the real worlds of policy, science, and human fallibility.

9.2.1 PLANNING

The box located in the upper left corner of Figure 9.1, labeled "planning (risk assessor/risk manager/interested parties dialogue)," is deceptively small. In fact, the planning dialogue is crucial to the outcome of an ecological risk assessment. The planning dialogue involves primarily the risk manager and the risk assessor.

Who are risk managers and risk assessors? A risk manager is usually a representative of an agency or organization that has authority to take action to mitigate ecological risks. A risk assessor is someone who is trained as a scientist and expert in the process of preparing risk assessments. Depending on the complexity of an ecological risk assessment, a risk assessor may work alone or as leader of a team of scientists.

Risk managers and risk assessors are expected to bring different but complementary perspectives to the risk-assessment process. The risk manager asks: What is the nature of the problem? What are the ecological entities that are threatened? What are the management goals and how will a risk assessment help with management decisions? What is the regulatory context of the assessment, e.g., Superfund site, national park, endangered species? What kinds of legal, policy, and societal considerations are involved? The risk assessor asks a different but complementary set of questions: What is the scale of the risk assessment? What are the critical ecological endpoints? What is the state of knowledge of the problem? What are the potential constraints in terms of limitations on available methods and data as well as limitations on expertise and time with which to conduct an assessment?

The dialogue between the risk manager and the risk assessor is supposed to result in specific planning products: clearly established management goals, agreement about the kinds of decisions that will need to be made in order to address the management goals, and agreement on the scope and complexity of the risk assessment. Other parties—the EPA framework refers to them as interested parties or stakeholders—are brought into the planning dialogue to the extent their participation is needed to establish management goals and implement management decisions. Stakeholder involvement may be minimal when management goals are set by statute, for example, in the registration of a new pesticide. When regulatory mandates are not clear-cut—in the protection of watersheds, for example—stakeholders may take on decisive roles in planning a risk assessment and partnering with government agencies in carrying out the management decisions the risk-assessment supports.

The term *management goal* deserves further definition. A management goal is defined as a "statement about the desired condition of an ecological value of concern." Examples are statements like "restore a stream," "prevent toxicity to wildlife," and "maintain a forest." Management goals embody values that society places on ecological entities, and they are essential as an impetus for initiating ecological risk assessments. However, management goals lack the conceptual precision a risk assessor needs in order to gather and analyze data related to risk. The necessary precision is provided by assessment endpoints.

9.2.2 PROBLEM FORMULATION

Formulating assessment endpoints is part of problem formulation, the step in the risk-assessment process that follows planning and that sometimes overlaps and interacts with the planning dialogue (Figure 9.1). To formulate useful assessment endpoints, it is generally necessary to integrate a broad range of background information about the ecological entity and the threat(s) it faces. Depending on the complexity of the risk scenario, the required background information may be culled from the scientific literature and assembled by one or two scientists, or it may require a team of scientists from several disciplines.

What is an assessment endpoint? An assessment endpoint is a statement of an ecological entity together with the attribute of the entity that is considered to be at risk. Examples of assessment endpoints are the trophic state (attribute) of a freshwater lake (ecological entity), the reproductive success (attribute) of peregrine falcons (ecological entity), and the species diversity (attribute) of an alpine meadow ecosystem (ecological entity). A key characteristic of an assessment endpoint is that it embodies a risk hypothesis that scientists can use to organize data. At the same time that it serves to organize data, a good assessment endpoint also expresses the management goal, i.e., the societal value, for which the risk assessment is being conducted. A good assessment endpoint builds a bridge from the decision to manage a particular risk, as articulated in the planning phase, to the scientific analysis that makes risk management possible. It sets the stage for investigating the effect of a stressor—chemical, physical, or biological—on an ecological receptor, be it a population, an ecosystem, or a planet. The assessment endpoint encapsulates the measurability of the stressor-receptor relationship that is at the core of the ecological risk assessment.

A second component of problem formulation is the construction of a conceptual model of stressor-receptor relationships (Figure 9.1). A diagram is helpful both in clarifying the relationships between stressors and receptors and in formulating assessment endpoints. A conceptual model of the effect of a physical stressor, logging, on an ecological receptor, a stream ecosystem, is presented in Figure 9.2. The primary stressor is the construction of logging roads, which give rise to a secondary stressor, increased siltation of streams. The primary effect of siltation is to smother benthic (bottom-dwelling) insects, and the secondary effect is a decreased abundance of insectivorous fish. Thus the conceptual diagram visualizes a set of stressor-receptor relationships that are hypothesized to connect logging with a diminished fishery. Importantly, each stressor-receptor relationship embodies a testable risk hypothesis. The diagram's predictions of increased stream siltation, decreased diversity and abundance of the benthic insect community, and decreased diversity and abundance of insectivorous fish are all testable. Each testable hypothesis is a potential assessment endpoint.

A second conceptual model showing the effect of a chemical stressor, lead, on populations of upland birds and their predators is diagrammed in Figure 9.3. Lead pellets that are ingested by upland birds or that are embedded in their flesh when they are wounded has the effect of either killing them as the result of lead toxicity

Ecological Risk Assessment

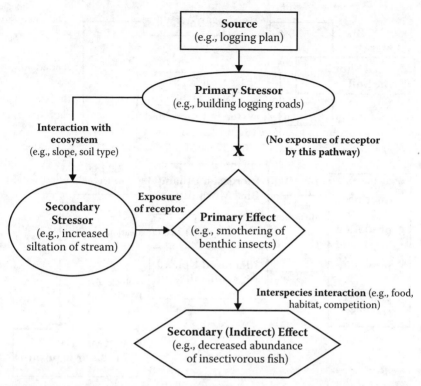

FIGURE 9.2 A conceptual model diagram of ecological stressors and receptors resulting from logging. Arrows indicate relationships between stressors and receptors. Each arrow represents a testable risk hypothesis. Note that an ecological receptor, e.g., benthic insects, can become an ecological stressor, e.g., decreased food supply for insectivorous fish. (Reprinted from U.S. Environmental Protection Agency, *Guidelines for Ecological Risk Assessment* [EPA/630/R-95/002F, http://cfpub.epa.gov/ncea/cfm/recordisplay.cfm?deid=12460, 1998], C-1.)

or debilitating them such that they reproduce less and become more prone to predation. Predators ingesting lead-poisoned birds are themselves exposed to lead poisoning. Upland bird populations decline, and so do populations of their predators. The arrows in the diagram suggest testable risk hypotheses. For example, increased bird morbidity due to ingestion of lead pellets might be tested by determining the effects of ingesting lead pellets on birds in controlled laboratory studies. Alternatively, it might be tested in field studies that observe the reproductive success of upland birds as a function of lead concentration in selected tissues.

The problem-formulation step of an ecological risk assessment concludes with the development of an analysis plan. The analysis plan identifies which assessment endpoints and risk hypotheses articulated by the conceptual diagram will be pursued in the next phase of the risk assessment: analysis (Figure 9.1). The decision as to which assessment endpoints to include in the risk assessment is guided by professional judgments on such questions such as the importance of a stressor-receptor relationship, its relationship to ecosystem structure, and the quality and availability of data to analyze it.

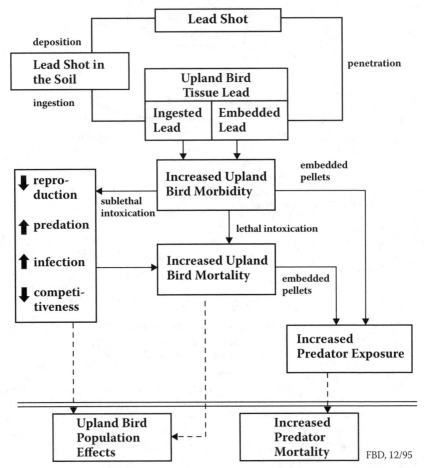

FIGURE 9.3 A conceptual model diagram of lead shot as an ecological stressor and upland bird populations and their predators as ecological receptors. Arrows track stresses on birds and predators. Note that an ecological receptor can become an ecological stressor, blurring the distinction between exposure and effect analyses. Thus, upland birds act as receptors with respect to lead shot and as stressors with respect to their predators. (Reprinted from U.S. Environmental Protection Agency, *Guidelines for Ecological Risk Assessment* [EPA/630/R-95/002F, http://cfpub.epa.gov/ncea/cfm/recordisplay.cfm?deid=12460, 1998], C-2.)

9.2.3 Analysis Phase

Like a human health risk assessment, an ecological risk assessment turns on analyses of exposure and effect. Exposure analyses are based on measures of exposure, and effects analyses are based on measures of effect; both types of measures are tentatively identified in the problem-formulation phase. There can be substantial interaction between exposure and effects analyses in natural systems, because a stressor may trigger a cascade of effects, some of which result in additional stressors and effects. For example, the effect of a logging road as a primary stressor is to create secondary stressors such as increased siltation of streams, and the primary effect of

Ecological Risk Assessment

siltation is to smother benthic insects, which reduces fish abundance (Figure 9.2). The potential for interaction between exposure analyses and effects analyses is indicated by the dashed line in Figure 9.1. Situated directly on top of the dashed line is a box labeled "measures of ecosystem and receptor characteristics." The position of this box is a reminder that the ecosystem plays a pivotal role in mediating exposure and effect; so does the behavior of the receptor itself.

Measures are crucial to a risk assessment because they make it possible to analyze risk. The EPA framework distinguishes three categories of measures: (a) measures of *effects*, which are measurable changes in a specified attribute of a receptor—or, alternatively, in its surrogate, if changes in the receptor chosen as the assessment endpoint cannot be measured directly; (b) measures of *exposure*, which measure the concentration and degree of contact or co-occurrence of a stressor with a receptor; and (c) measures of *ecosystem characteristics* that affect the behavior and distribution of receptors and stressors.

The different categories of measures and their relationship to assessment endpoints and management goals are illustrated by the salmon fishery in the Pacific Northwest:

> *Management goal*: A viable, self-sustaining coho salmon population that supports a subsistence fishery for Native American people and a sport fishery for anglers
> *Assessment endpoint*: Coho salmon breeding success, fry survival, and adult return rates
> *Measures of effects*: Effects of low dissolved-oxygen concentrations on fish eggs and fry; effects of obstacles such as dams on adult fish behavior; and effects of bottom sediment in streams on spawning behavior and egg survival
> *Measures of exposure*: Number of hydroelectric dams and associated obstructions to fish passage; concentrations of toxic chemicals in water, stream sediment, and fish tissue; concentrations of nutrients and dissolved oxygen in ambient waters; and riparian (stream bank) cover, sediment loading, and water temperature
> *Measures of ecosystem and receptor characteristics*: Water temperature, water velocity, and physical obstructions; abundance and distribution of suitable pebble substrate for breeding in streams; abundance and distribution of suitable food sources for fry; feeding, resting, and breeding behavior of adult fish; and natural reproduction, growth, and mortality rates

The analysis step need not include all available measures. However, an analysis does need to include enough measures to assess the risks that threaten attainment of the management goal.

Let us take a closer look at the analysis of exposure to a toxic chemical. If the ecological receptor is a population of fish, such as coho salmon, a useful measure of exposure is the concentration of the toxic chemical in water. If the ecological receptor is a population of birds, exposure analysis is performed in much the same way as the estimate of chronic daily intake in a human health risk assessment (Chapter 8). The measure of interest is the average daily dose (ADD) in units of milligrams of toxic chemical per kilogram of body weight. If birds ingest the toxic chemical with their food, then the average daily dose is the product of the average contaminant concentration in a food type or prey species (C), the fraction of the food type that comes from the contaminated area (FR), and the normal ingestion rate (NIR) of the food type in units of kilograms food per kilogram body weight per day, then ADD is defined as shown in Equation (9.1):

$$ADD = C \times FR \times NIR \qquad (9.1)$$

The ADD calculation is repeated for each food type, and the products are summed to obtain the average daily dose received by the bird population from all food types.

The calculation of an average daily dose for birds suggests the potential for interactions between measures of exposure, on the one hand, and measures of ecosystem and receptor characteristics, on the other (refer to the box on the dotted line in Figure 9.1). For example, the average contaminant concentration, C, depends on the receptor's food preferences. It also depends on the characteristics of the food web and the degree to which the toxic chemical may be bioconcentrated and biomagnified. Ecosystem characteristics can also affect the abundance of various food types. These examples illustrate the point that the particular characteristics of a receptor and its ecosystem may contribute to exposure, and that such characteristics need to be taken into account along with a chemical's environmental fate and transport when analyzing exposure.

Another approach to analyzing chemical exposure is analogous to human biomonitoring: Capture members of the receptor population—fish or birds, in the two proposed examples—and measure the concentration of chemical in their tissues. As with human biomonitoring, inferring toxic effect frequencies from tissue concentrations is problematic, and for the same reason: Effect frequency is generally studied as a function of the external dose of a chemical to which a receptor is exposed—e.g., the concentration of the chemical in water, food, or air—and not as a function of the concentration of chemical that ends up in the receptor's tissues. Nevertheless, tissue concentrations can provide valuable information about exposure, even if they cannot be compared to the results of dose–effect studies in laboratory animals.

Like measures of exposure, measures of effect are crucial to the analysis of risk. The choice of measure depends on the management goal and the assessment endpoint. If the management goal is the health of a wildlife population and the assessment endpoint is its reproductive success, measures of effect may include mating and feeding behavior, numbers of viable offspring, and population size. If the management goal is the health of an aquatic ecosystem such as a lake or a stream, the assessment endpoint may be trophic state or productivity, i.e., the ability to support aquatic life. An ecosystem attribute such as trophic state is impossible to measure directly, and a surrogate assessment endpoint must be chosen to represent it. For example, the abundance of algae is a common surrogate for a lake's trophic state. The amount of chlorophyll a in lake water is a function of algae abundance, and chlorophyll a concentration can therefore be used as an indirect measure of effect on the lake's trophic state due to stressors such as phosphorus and nitrogen nutrients. In the case of a stream ecosystem, a surrogate for productivity as an assessment endpoint might be the health of the benthic

Ecological Risk Assessment

insect community; thus measures of effect might include the relative abundance of pollution-sensitive and pollution-tolerant families of insects.

Measures of effect may be obtained from published studies on the ecological entity of concern. Effects may also be estimated by extrapolating from studies of other receptors that are subject to comparable exposure scenarios. When the stressors are chemicals, the EPA's online ECOTOX database provides a comprehensive source of information on the effects of thousands of toxic chemicals on hundreds of species of animals and plants. Data gaps may be filled by performing field or laboratory studies as part of the ecological risk assessment.

The output of the analysis phase of an ecological risk assessment consists of an exposure profile and a stressor-response profile (Figure 9.1). The profiles include estimates of uncertainty, which, as in human health risk assessments, are a constant and essential ingredient of ecological risk assessments. The exposure and effect profiles, together with their associated uncertainty estimates, set the stage for the final phase of the risk-assessment process.

9.2.4 Risk Characterization

As in a human health risk assessment, an ecological risk assessment concludes with a narrative characterization of risk (Figure 9.1). The evidence marshaled in the exposure and effect analyses provides the basis for estimating the risk to the ecological value identified in the management goal. The risk estimate includes quantitative assessments when possible. For example, the results of studies on the effect of lead shot on ring-necked pheasant under controlled laboratory conditions may allow an estimate of the mortality of upland birds exposed to lead shot in the field (Figure 9.3). Risk estimates also include qualitative assessments. For example, siltation of streams caused by logging roads is predicted to result in decreases in insectivorous fish populations (Figure 9.2), but numbers cannot be put on the size of the decreases. Quantitative and qualitative estimates of risk are summarized in a final narrative statement that describes the risk to the ecological value articulated in the management goal and the uncertainties associated with risk estimates.

9.2.5 Risk Communication and Adaptive Management

The risk characterization authored by the risk assessor, who is a scientist, becomes the basis for communicating with the risk manager, who generally is not a scientist, about risk to the ecological value of concern. Communicating risk to the risk manager—who may not be versed in all of the analyses that go into a risk assessment—requires considerable skill on the part of the risk assessor. Good communication from risk assessor to risk manager is essential to translating the risk assessment's conclusions into actions to protect the threatened resource. The risk manager, in turn, communicates the risk to stakeholders, both those who will not and those who will become active participants in protecting the resource. The risk manager is responsible for devising and coordinating a management plan. The plan must mitigate the risks described in the risk assessment, and at the same time, it must remain fluid enough to adapt to possible changes in the understanding of the risk as the results of implementing risk-management activities are monitored and as scientific studies generate new information (Figure 9.1)

9.3 ENVIRONMENTAL IMPACT STATEMENT

An environmental impact statement (EIS) is a form of ecological risk assessment. It differs in that its focus is broader, encompassing cultural and economic resources in addition to ecological entities that may be threatened. Environmental impact statements are mandated under the National Environmental Policy Act (NEPA) and similar state laws as a prerequisite for many types of development. These laws generally require that environmental impacts be avoided, minimized, rectified, and/or compensated. The EIS process includes a public comment period lasting 30 to 90 days when stakeholders, including any member of the general public, may voice their opinions and concerns. The public comment period is particularly important because it ensures that agencies are watched, by each other and by members of the public, as they craft their decisions. The government agency conducting the EIS is required by law to respond to all comments before finalizing the EIS and allowing a development project to go forward. Possible responses include modifying the project to reduce an impact; explaining that an impact is less severe than a comment suggests; or acknowledging that an impact will occur but that the harm is outweighed by the economic benefits the project will provide.

Enforcing environmental protections under the EIS process relies primarily on case law, and protections are strongest when legal precedents are clear. A weakness of the case-by-case approach is that protections tend to focus on site-specific harm rather than on larger processes that operate on a landscape scale and over long periods of time. Laws in other countries have begun addressing toxic chemical risk in larger spatial and temporal contexts. In the final chapter we examine some of the innovative legislative strategies emerging from Canada and the European Union.

STUDY QUESTIONS

1. Ecological risk assessments and human health risk assessments follow similar processes, but they differ in important ways. Discuss three differences that you think are significant.
2. Choose three ecological values that are important to you and articulate management goals for them.
3. Craft assessment endpoints for each of the management goals that you articulated in Question 2.
4. What is a conceptual model diagram, and why is it useful in an ecological risk assessment?
5. Name the three major classes of ecological stressors and give at least one example of each.
6. What is a measure of exposure? Identify at least one measure of exposure for each of the three classes of stressors in Question 5.
7. What is a measure of effect? Name three ecological receptors and at least one measure of effect for each.

8. What is a measure of ecosystem characteristics? Name three ecosystem characteristics and one measure for each.
9. As a thought experiment, design an ecological risk assessment to support one of the management goals that you articulated in Question 2. Beginning with the assessment endpoint formulated in Question 3, draw a conceptual model diagram of the relationships between stressor(s) and receptor(s). Make a list of measures of exposure to the stressor(s) and measures of effect on the receptor(s). Make a separate list of ecosystem and receptor characteristics that could modify exposure and/or the effect of exposure on the receptor, and identify a measure for each characteristic. As the final step in this thought experiment, set up analyses of exposure and effect by designing studies—field studies, laboratory studies, or both—that provide data for the measures of exposure, effect, and ecosystem and receptor characteristics that you have identified.

ANSWERS TO STUDY QUESTIONS

1. (a) An ecological risk assessment begins with the articulation of one of many possible ecological values to be protected. The value to be protected by human health risk assessments is simple: public health. (b) As they cascade through an ecosystem, stresses can become effects, and the effects created by stressors can in turn become stressors themselves. In a human health risk assessment, the stressor is always the disease agent and the effect is always the disease. (c) The products of a human health risk assessment generally include one or more quantitative indices of risk, e.g., the level of exposure expressed as the hazard index (HI), the risk of cancer expressed as a probability, or the therapeutic index of a drug. While an ecological risk assessment may employ a wide range of numerical measures, its products tend to be descriptions of risk rather than numerical indices of risk. Notwithstanding the differences in their products, the levels of uncertainty associated with the two forms of risk assessment are roughly comparable.
2. Answer depends on reader's initiative.
3. Answer depends on reader's initiative.
4. A conceptual model diagram shows the relationships between stressors and receptors in the context of the ecosystem where the stresses occur. Because these relationships are often complicated, diagramming them is helpful in formulating assessment endpoints and in identifying specific measures for use in exposure and effect analyses.
5. (a) Physical stressors, for example, degradation of habitat (deforestation, rise in temperature, dams). (b) Chemical stressors, for example, greenhouse gases, endocrine disruptors, pesticides, metals. (c) Biological stressors, for example, invasive species (Dutch elm disease, purple loosestrife, zebra mussels).
6. A measure of exposure is a measure of the degree to which a receptor is exposed to a stressor. (a) Measures of physical stressors: fragmentation of forest habitat, degrees of temperature rise, number of dams. (b) Measures of chemical stressors: concentrations in environmental media, co-occurrence with

receptors. (c) Measures of biological stressors: geographic range, reproductive success.
7. A measure of effect is a measure of the degree to which an ecological receptor is impacted by a stressor. Three examples: (a) The receptor is a freshwater lake, the stressor is the nutrient phosphorus, and the effect is increased productivity of the lake, which could be measured as the abundance of green algae. Instead of algae, a surrogate measure, the concentration of chlorophyll a, is commonly used. (b) The receptor is bald eagles in the Great Lakes region, the stressor is the pesticide DDT and its degradation products such as DDE, and the effect is a decrease in bald eagle populations. Measures of effect include census counts of threatened populations, number of nests, and hatching success. (c) The ecological receptor is the climate of planet Earth, the stressor is the greenhouse gas carbon dioxide, and the effect is changes in rainfall patterns. Measures of effect include inches of rain per year, frequency of rain, amount of rain per rainfall event, distance of groundwater table from the ground surface, flows in streams, and levels of lakes.
8. An ecosystem characteristic is an attribute that is believed to be integral to the structure and function of an ecosystem. Identifying ecosystem characteristics is, of course, dependent on the extent to which an ecosystem is understood. Examples: (a) A stream is an aquatic ecosystem; it may also contain several distinct ecosystems along its length. Measurable ecosystem characteristics include water depth, current velocity, temperature, and oxygenation. Stream bank vegetation is part of the stream ecosystem, because it helps regulate temperature by providing shade. The extent of stream bank vegetation can also be measured. (b) A desert ecosystem could be characterized, in part, by such measures as annual rainfall, temperature, humidity, and the plant and animal communities it supports. (c) Measures of the characteristics of a forest ecosystem might include species of trees, their ages, and the insects and birds that inhabit them.
9. Answer depends on reader.

REFERENCES

U.S. EPA. 1994. Review of ecological assessment case studies from a risk assessment perspective. Vol. 2. Environmental Protection Agency, EPA/630/R-94/003. http://cfpub.epa.gov/ncea/cfm/recordisplay.cfm?deid=30819

U.S. EPA. 1998. Guidelines for ecological risk assessment. Environmental Protection Agency, EPA/630/R-95/002F. http://cfpub.epa.gov/ncea/cfm/recordisplay.cfm?deid=12460

SUGGESTED READING

Gerba, C. P. 1996. Risk assessment. In *Pollution science*, ed. I. L. Pepper, C. P. Gerba, and M. L. Brusseau, pp 345–364. New York: Academic Press.

10 Managing Chemical Risk in North America and Europe

10.1 INTRODUCTION

This book began by noting that toxic chemicals are double-edged swords, offering many benefits to society but also presenting dangers to human health and the environment. The dangers can be evaluated through the process of risk assessment (Chapters 8, 9), assuming that data on toxicity and exposure are available. Risk assessment provides a rational basis for prioritizing risks and taking steps to protect public health and natural resources. Because risk assessment—if given a chance and used appropriately—can and does work, the next question is, "How can the insights of risk assessment be translated into the fabric of everyday life, where exposures to toxic chemicals occur?" In other words, how does society, on a day-to-day basis, protect itself from the intrinsic risks of toxic chemicals? Before discussing specific strategies, it is worth recalling the kinds of problems toxic chemicals can create if they are not managed well.

10.2 COSTS OF TOXIC CHEMICALS TO SOCIETY

The potential cost to society of everyday uses of toxic chemicals can be divided into two broad categories: Undermining public health and damage to the environment. The two categories are considered separately, although they are closely related.

10.2.1 UNDERMINING HUMAN HEALTH

Human health may be impacted directly or indirectly by toxic chemicals. Direct health impacts can result from exposure to toxic chemicals in the food we eat, the water we drink, and the air we breathe as well as from products that contact our skin such as cosmetics. Food might contain pesticide residues used by farms and agribusinesses to prevent damage to crops. Toxic chemicals might leach into food from the packaging used to store it and from the cookware used to prepare it. Drinking water might become contaminated when industries discharge hazardous chemical by-products of manufacturing processes into streams or dispose of them in landfills that eventually leak and contaminate groundwater aquifers. Other industrial by-products, such as smog and particulate matter, can contaminate air and cause or exacerbate respiratory diseases. If direct exposure to a toxic chemical in food, water, and/or air exceeds its threshold dose, the chemical's toxicity manifests itself in the form of specific toxic effects, beginning with the most vulnerable members of the exposed population. In other words, people start getting sick, and those with chronic medical conditions, the elderly, and the young tend to get sick first. Chemicals like genotoxic carcinogens that may lack thresholds (or may not—the science is as yet inconclusive [Chapter 8]) will also cause dose-dependent increases in morbidity and mortality, starting with the most vulnerable individuals. Exposure scenarios often involve multiple chemicals, as suggested by biomonitoring studies (Chapter 4).

In principle, toxic effects resulting from exposure to chemical mixtures may be additive (equal to the sum of chemicals' individual toxicities), synergistic (greater than the sum of individual toxicities), or inhibitory (less than the sum of individual toxicities). However, little is known about the actual effects of specific mixtures of toxic chemicals on health. Notwithstanding gaps in our understanding of risk, the general validity of the dose–effect relationship tells us unequivocally that the greater society's exposure to toxic chemicals, whether short-term or long-term, alone or in mixtures, the higher is the incidence of disease.

Hazardous chemicals may also undermine human health indirectly when they interfere with natural systems and resources that support health. One example is food security. Global warming due to the release of greenhouse gases from burning fossil fuels to power the global economy is predicted to cause regional changes in the hydrologic cycle (Figure 2.3) that will result in increased frequencies of droughts and floods, which in turn will damage agriculture and reduce food supplies (Houghton 2004). Another example is the increased atmospheric concentration of the predominant greenhouse gas, carbon dioxide, which is predicted to cause acidification of the oceans. Ocean acidification could affect marine life, beginning with organisms like coral reefs, whose shells depend on pH-sensitive calcification reactions and possibly leading to reduced quantities of fish and other food sources for human populations (USGS 2008). Besides threatening food security, hazardous chemicals can undermine natural systems that protect human populations from disease. For example, the "hole" in stratospheric ozone that is caused by chlorofluorocarbons (CFCs) results in a greater intensity of ultraviolet light reaching the Earth's surface and a higher incidence of skin cancer. Increasing temperatures in a warming climate make it possible for disease vectors, such as mosquitoes that play host to the malaria parasite, to spread to human populations that were previously outside their range.

10.2.2 Damage to the Environment

Toxic chemicals may impact ecological receptors directly by poisoning them or indirectly by initiating a cascade of stresses (Chapter 9). Impacts on ecological receptors are often measured in terms of the failure of exposed populations to reproduce. One example is the presence of endocrine-disrupting chemicals that enter a river from an industrial facility or from normal household waste that passes unchanged through a sewage treatment plant, interfering with normal sexual development and resulting in intersex fish (fish with sexual characteristics intermediate between female and male) and reproductive failure. Another example is DDE, a metabolite of the insecticide DDT, generated by soil bacteria. DDE causes eggshell thinning and hence reproductive failure in birds. A third example of direct poisoning is the "acid rain" that can result from sulfur dioxide emissions from coal-burning power plants. Prior to passage of the Clean Air Act, acid rain lowered the pH in hundreds of lakes in New York State's Adirondack Park, which lies downwind from coal-burning power plants in the midwestern United States. Many lakes were no longer able to support aquatic life.

Fertilizer applied to cropland offers an example of indirect effects that chemicals can have on the environment. Fertilizers containing phosphorus and nitrogen are applied to crops to enhance their growth. A portion of the fertilizer is not utilized by crop plants, and this excess reaches rivers and other surface water bodies by percolating into groundwater as well as by running off the ground surface with rain or snowmelt. The phosphorus and nitrogen move with the river (an example of advective transport, Chapter 2), ending up in lakes and seas where they fertilize the growth of algae and other vegetation. When aquatic plants die, they are broken down by decomposer organisms (Figure 2.2), a process that depends on oxygen from the atmosphere dissolving in water (Henry's law, Chapter 2). Excessive quantities of dead algae and other plants can result in the depletion of dissolved oxygen by microbial decomposer organisms and, ultimately, the death of fish and other aquatic life. The Gulf of Mexico now contains a dead zone measuring hundreds of square miles near the mouth of the Mississippi River. Over 400 similar dead zones have been identified in coastal waters around the world (Vaccari 2009).

While chemicals can harm the environment, it is important to keep their impacts in perspective. Ecological receptors are harmed by two other types of stressors: (a) the physical destruction of their habitat resulting from the expansion and development of cropland, suburbs, and cities and (b) the introduction of invasive species, facilitated by human mobility, that destroy or outcompete native species. It seems possible that humanity may be causing—largely unwittingly—extinctions of plant and animal species on a scale that rivals extinction events in past geologic epochs. Toxic chemicals are a significant anthropogenic stressor, but they are by no means the only stressor of natural systems today (Kolbert 2009).

10.3 CORE CONCEPTS OF RISK MANAGEMENT

How does contemporary society manage the downsides of the chemicals we depend on to sustain our global civilization? Managing chemical risks is a dynamic area of law and governance, and one that is evolving rapidly with increasing awareness of the risks associated with toxic chemicals. Promising new strategies build on earlier efforts to rein in risks. While diverse in detail, all management strategies have a few core concepts in common, as described in the following four subsections.

10.3.1 CONTROLLING EXPOSURE

The most effective way to reduce risk is to reduce exposure. Risk is a function of toxicity and exposure (Chapter 1). Exposure can be controlled, while toxicity cannot; toxicity does, however, indicate the degree to which exposure needs to be controlled. The more toxic a chemical, the lower the doses to which humans and other species may be exposed without harm. (It should be noted that toxicity is a function of a chemical's toxic effect and its potency (Chapter 5); therefore it would be more precise to say the greater the potency of a toxic chemical, the lower is the acceptable dose.)

10.3.2 Dealing with Uncertainty

Risk management is associated with scientific uncertainty. Uncertainty may be reduced by acquiring more data, but it can never be eliminated completely. Depending on the degree of uncertainty involved, risk-management decisions may need to be iterative or adaptive, explicitly evolving in response to new information and understanding of specific risks (see Figure 9.1). Lack of scientific certainty is not in itself a sufficient reason for postponing important risk-management decisions. If society were to wait for absolute scientific certainty, very few risks—of any kind—would be regulated.

10.3.3 Navigating the Political Process

Risk management is embodied in laws, and as such it is part of the normal political process, involving economic and worldview factors in addition to science (Figure 1.2). The effectiveness of risk management may ultimately depend as much—if not more—on the quality of the political process than it does on the science that goes into a good risk assessment.

10.3.4 Balancing Uncertainty, Proof, and Precaution

Given scientific uncertainty, risk managers are forced to strike a balance between cautiously preempting harm to human health and the environment, on the one hand, and, on the other hand, waiting for more persuasive proof of harm before imposing a regulatory burden on businesses. The preemptive side of this continuous balancing act was first articulated in Germany in the 1970s as the "precautionary principle" (German: *Vorsorgeprinzip*). It is expressed colloquially as "better safe than sorry." Applied to toxic chemical risk, the precautionary principle generally translates to a willingness, in the face of considerable scientific uncertainty, to regulate exposure more aggressively in order to forestall what are anticipated to be the potentially unacceptable effects of a chemical's toxicity. The business side of the balance emphasizes the importance of regulatory flexibility in order to maintain economic growth and encourage technological innovation.

10.4 GENERAL STRATEGIES FOR MANAGING TOXIC CHEMICAL RISK

10.4.1 Sorting Chemicals

Managing toxic chemical risk begins with deciding which chemicals are sufficiently toxic to warrant regulation and which are not. This requires, first, that chemicals be characterized with respect to their toxicity (Chapter 5), and second, that criteria be established for sorting chemicals into those that are sufficiently toxic to require regulation and those that are, for all intents and purposes, nontoxic and do not require regulation.

Sorting chemicals into toxic and nontoxic is the first step in regulating toxic chemical risk. Incredibly, at the time of this writing (November 2009), this first step has not yet been accomplished in the United States or any other country, with the sole exception of Canada. In the United States, less than 10% of approximately 84,000 chemicals reported to be in commerce have been characterized with respect to their toxicity, and there are no plans to screen the other 90%. The situation had been similar in Europe; however, the European Union (EU) recently passed legislation called REACH (Registration, Evaluation, Authorization and Restriction of Chemicals) that mandates toxicity assessments of all chemicals that are used in products sold in the EU. REACH is designed to profile the toxicity of each and every chemical in commerce, ultimately plugging the huge data gaps left by earlier laws and placing chemical risk assessments on a firmer footing. The new European law, described in greater detail in Section 10.5.3, represents a significant regulatory advance in that it establishes explicit rules while providing extensive administrative support for businesses to comply with the law.

Sorting chemicals into toxic and nontoxic is a necessary first step, but it is not sufficient for regulatory purposes. Toxic chemicals are extremely diverse with respect to their toxicities, their physical-chemical properties, and the ways they are incorporated into commercial products. Diverse product uses add to the complexity. One of the great challenges of managing toxic chemical risk is to devise a regulatory scheme that simplifies the universe of toxic chemicals enough to make it manageable under real-world conditions yet avoids an oversimplified, one-size-fits-all approach that could stifle business and industry. There need to be consistent and transparent "rules of the road" for using toxic chemicals, yet the rules need to be flexible enough for waivers to be granted in response to reasonable requests from entrepreneurs. The question then becomes, "How might toxic chemical risk be managed effectively within the existing economic and political framework of industrialized nation-states?" It is worth noting that some, including some business leaders, argue that rules can benefit entrepreneurial activity by reducing uncertainties in business models.

10.4.2 Limiting the Use of Toxic Chemicals

One management tool available to governments is to pass laws restricting the use of a chemical or class of chemicals whose hazards are judged to outweigh their benefits under certain conditions of use. The limitations may vary, depending on the toxicity of the chemical and how it is used. For example, the Toxic Substances Control Act (TSCA) (1976) banned polychlorinated biphenyls (PCBs) from most commercial uses in the United States as a result of mounting evidence of PCB toxicity. To reverse the destruction of stratospheric ozone, the Montreal Protocol (1987) mandated a ban on the use of chlorofluorocarbons and their replacement by other chemicals as refrigerants and aerosols. In an effort to manage the burgeoning problem of electrical and electronic waste, the European Union in 2006 issued a directive known as the restriction on the use of certain hazardous substances in electrical and electronic equipment (RoHS). RoHS limits the quantities of six chemicals, including four metals (lead, mercury, cadmium, and hexavalent chromium) and two classes of flame retardants (polybrominated biphenyls and polybrominated biphenyl ether) that may be used in electrical and electronic equipment. Typically, laws have restricted the quantities of toxic chemicals and the conditions under which they may be used rather than banning the chemicals outright.

10.4.3 LIFE-CYCLE ASSESSMENT

A life-cycle assessment (LCA) chronicles the life of a commercial product from manufacture to disposal ("cradle to grave") in order to estimate its true cost and, at the same time, to discover how the product's life cycle might be modified to increase profitability within the framework of existing laws and regulations. The familiar admonitions to "reduce, reuse, recycle" seek to apply LCA concepts to both ends of products' life cycles by minimizing both inputs of natural resources and the amounts of waste products. When applied to toxic chemicals, an LCA examines the quantities of toxics used in manufacturing a product, the exposure of workers who make the product and of consumers who use it, and the environmental fate and transport of toxic chemicals after the product is disposed of. Pollution prevention is based on the life-cycle-assessment approach and has been embraced by many governments.

Environmental laws regulating the disposal of hazardous chemical waste may also be informed by the LCA concept. The Waste in Electrical and Electronic Equipment (WEEE) directive in the European Union is closely linked to the RoHS directive restricting the use of specific chemicals in electrical products. The Resource Conservation and Recovery Act (RCRA) in the United States complements other environmental laws that regulate toxic chemicals by regulating the disposal of the hazardous chemical by-products of industrial processes. A significant difference between RCRA and WEEE is that WEEE is linked to pollution prevention laws such as RoHS, while RCRA is more of an "end of the pipe" law.

10.5 ENVIRONMENTAL LAWS IN NORTH AMERICA AND EUROPE

The United States was among the first countries, beginning in the 1970s, to pass laws regulating the use of toxic chemicals with the goal of protecting human health and the environment (Table 4.1). Many industrialized nations followed the lead of the United States, modeling their laws on ours. A generation later, both Canada and the European Union have moved to strengthen their environmental laws, going beyond the regulatory framework established by the United States and charting a new, more precautionary course in managing toxic chemical risk.

10.5.1 THE TOXIC SUBSTANCES CONTROL ACT

Before discussing the regulatory initiatives underway in Canada and the European Union, let us consider first the statutory framework that governs the regulation of toxic chemicals in the United States. While a number of laws regulate toxic chemicals (Table 4.1), the centerpiece of toxics regulation is the Toxic Substances Control Act (TSCA) (Cornell University Law School 2009; U.S. EPA 2009). Under TSCA, businesses are required to report to EPA any chemical that is manufactured or imported in quantities greater than 10,000 pounds (4.54 tons) at any single site. The report contains contact information for the company, the name of the chemical, and the amount for each site above 10,000 pounds. At the time the law was passed, businesses were not required to provide information on toxicity or exposure for chemicals reported to EPA, approximately 62,000 chemicals in all. However, in order to manufacture or import a "new" (i.e., post-TSCA) chemical, businesses are required to provide EPA with a premanufacture notification (PNA), which must include all available information on the chemical's health and ecological effects, physical and chemical properties, and environmental fate characteristics.

All chemicals reported to EPA are placed on the TSCA Inventory, which currently contains approximately 84,000 chemicals. The EPA can ban or restrict the use of any chemical, whether pre- or post-TSCA, if it "presents an unreasonable risk of injury to health or the environment" (Subchapter I, Paragraph 2605(a)). At the same time, the stated intent of TSCA is not to "impede unduly or create unnecessary economic barriers to technological innovation" (Subchapter I, Paragraph 2601 (b) (3)). In practice, the bar against stifling technological innovation has turned out to be so high that an EPA decision to ban the use of asbestos, a known human carcinogen, was struck down in federal court in 1989 in a lawsuit brought by industry. The asbestos case made it clear that under TSCA, federal and state governments have little choice but to rely to a significant degree on the voluntary cooperation of business and industry (Schapiro 2007).

As mentioned earlier, one of the core issues in regulating toxic chemical risk is that laws must strike a balance between preempting harm through government oversight and creating a regulatory environment where businesses can prosper. The balance that TSCA strikes relies to a significant extent on the good faith of businesses, first, in reporting the toxicities of chemicals in their products, and second, in accepting competitive limitations that could potentially result from restricting their use of those chemicals. While the intent of the law is protective, many critics believe that the policies and regulations that have grown out of TSCA have proven inadequate to the task of protecting human health and the environment.

10.5.2 Canadian Environmental Protection Act (CEPA)

According to the Canadian government's Web site, CEPA is informed by the principles of sustainable development, the precautionary principle, science-based decision making, and an ecosystem approach, among others (http://www.ec.gc.ca/CEPARegistry/the_act/guide04/s3.cfm). Similar to TSCA, CEPA requires that chemical substances manufactured or imported into Canada be identified and reported to the government, which places them on the Domestic Substances List (DSL), the Canadian counterpart of the TSCA Inventory. The threshold for reporting is much lower in Canada, 100 kilograms (roughly 200 pounds). While TSCA evaluates chemicals on a case-by-case basis, the Canadian law requires that all substances be categorized (screened) using standardized criteria for evaluating impacts on human health and the environment. The Domestic Substances List was compiled in the early 1990s, and new chemicals are added continuously. There are currently about 24,500 substances on the DSL.

Screening of substances on the DSL is based on the definition of "toxic" in the Canadian Environmental Protection Act: "A substance is toxic if it is entering or may enter the environment in a quantity or concentration or under conditions that: a) have or may have an immediate or long-term harmful effect on the environment or its biological diversity; b) constitute or may constitute a danger to the environment on which life depends; or c) constitute or may constitute a danger in Canada to human life or health." Thus CEPA frames toxicity in terms of impacts on the environment as well as the traditional definitions based on health effects, e.g., neurotoxicity, birth defects, and cancer. The general definition in the statute is translated into specific criteria that are used to decide whether a substance is, indeed, toxic within the meaning of the law. These criteria include a substance's persistence in the environment, its capacity to undergo bioaccumulation, its inherent toxicity, and the degree to which humans may be exposed (http://www.ec.gc.ca/toxics/TSMP/en/criteria.cfm).

The screening of chemicals on the DSL is administered jointly by the Ministries of Health and the Environment. Chemicals that meet one or more of the screening criteria are considered candidates for classification as "toxic" as defined by CEPA and are placed in a pool for further study. When screening of all of the substances on the DSL was completed several years ago, approximately 4,000 substances were found to meet one or more of the criteria for toxicity as defined by CEPA. Substances believed to pose the greatest risks are moved to a Priority Substances List for in-depth risk assessment. If determined to be toxic within the meaning of the Canadian Environmental Protection Act, a chemical is placed on Canada's Toxic Substances List, also referred to as Schedule 1. To date, some 200 chemicals have been classified as "CEPA toxic" and placed on Schedule 1.

How does the Canadian government manage risks from "CEPA toxic" chemicals? Development of concrete management plans is ongoing. Beginning in 2006, the Canadian government published a timetable (referred to as "The Challenge") to develop plans for managing Schedule 1 chemicals. The Challenge divides the 200 substances currently on Schedule 1 into a number of smaller groups, called batches, containing roughly a dozen chemicals each. Batches are being addressed sequentially. Industry and other stakeholders are invited to provide additional information on the chemicals in each batch in an effort to refine risk assessments and develop best practices for risk management. Risk-management plans are currently being released for a new batch of Schedule 1 chemicals every three months. Interested readers can follow the unfolding Challenge process on the Internet (Government of Canada 2009).

While practical details vary from chemical to chemical, government policy under CEPA provides two broad management tracks for CEPA toxic (Schedule 1) chemicals. Track 1 applies to chemicals that meet all of the standardized criteria for toxicity. Track 1 toxics are managed by means of "virtual elimination," which is defined as "the reduction of releases to the environment of the most dangerous toxic substances to a level below which these releases cannot be accurately measured." As of February 2009, one chemical and one group of chemicals had been targeted for virtual elimination: hexachlorobutadiene (C_4Cl_6) and perfluorooctane sulfonate (PFOS) and its salts. Track 2 toxics fulfill some but not all of the standard criteria for toxicity. Risk from Track 2 toxics is managed in consultation with industry by modifying products' life cycles to prevent or minimize releases to the environment. It appears that most of the chemicals on Canada's Toxic Substances List will be managed as Track 2 toxics.

In summary, CEPA 1999 goes well beyond TSCA by mandating that all chemicals in commerce in Canada be screened using standardized criteria for toxicity, by enlarging the definition of toxicity to include environmental persistence and bioaccumulation, and by systematically prioritizing chemicals for management, including total or near-total bans on the most toxic substances (virtual elimination). The Canadian law is a promising new initiative in the area of toxics regulation, but it remains to be seen how well it is able in practice to benefit the environment and human health without putting Canadian business at a competitive disadvantage.

The case of bisphenol A (BPA), an endocrine disruptor, suggests how CEPA 1999, and specifically its incorporation of the precautionary principle, might affect human health. BPA is not on Schedule 1; however, concerns have been raised regarding its possible impacts on human development up to the age of 18 months. The Canadian government proposed regulations in June 2009 to prohibit the advertisement, sale, and importation of polycarbonate plastic baby bottles that contain BPA. The risk assessment concluded that BPA leaches out of the bottles and into baby formula, but at concentrations below those that cause adverse health effects in infants. Nevertheless, the government decided that BPA concentrations were sufficiently close to the threshold of toxicity, and the degree of uncertainty in the risk assessment was sufficiently large to warrant banning BPA-containing bottles as a precautionary measure. Using essentially the same risk assessment, other countries, including the United States, have decided not to ban BPA-containing baby bottles. Given the scientific uncertainties involved, it is virtually impossible to say who is right. However, the Canadian government's decision to err on the side of precaution is consistent with CEPA 1999.

10.5.3 REACH

The European Parliament and the Council of the European Union passed a far-reaching law called the directive on the Registration, Evaluation, Authorization and Restriction of Chemicals (REACH), which took effect on June 1, 2007. REACH was years in the making, and it was reportedly opposed by the United States under the administration of President George W. Bush because it was perceived as conflicting with U.S. business interests (Schapiro 2007). As its name implies, REACH mandates that each and every chemical in commerce in the European Union, both those that originate in the EU and those that are imported, be registered and evaluated with respect to risks to human health and the environment; that the use of high-risk chemicals be subject to specific restrictions, including tonnage limitations and bans; and that restrictions on high-risk chemicals be waived only on specific authorization by the EU government. REACH places significant responsibility on industry to manage toxic chemical risk. Unlike TSCA, which makes risk management by industry largely a voluntary matter in the United States, REACH contains explicit rules that businesses must follow. The law creates a European Chemical Agency (ECHA) with headquarters in Helsinki, Finland, to support industry in complying with rules for registering chemicals and managing risks and to provide expert guidance to the EU government on the restriction and authorization of hazardous chemicals. ECHA makes extensive use of the Internet to communicate with businesses and the public. While English is used primarily, ECHA disseminates documents in 22 languages.

REACH, like CEPA (and TSCA), requires that all existing chemicals be inventoried. REACH refers to the inventory process as "registration." The legal responsibility for registering chemicals falls squarely on producers and importers. Registration is based on individual chemicals, regardless of whether a chemical is used alone, in a mixture, or in a product. Chemicals that are manufactured or imported into the European Union in quantities greater than 1 ton/year per producer or importer are subject to registration. A few substances are exempted because they are covered by other laws, are used in research, or are judged to be harmless.

To register a chemical under REACH, a producer or importer must submit a technical dossier containing basic information about the chemical, including any risks it may present to human health and the environment. The technical dossier describes: the intrinsic properties of the chemical that are relevant to evaluating its fate and transport in the environment; whether the chemical is known to be dangerous to human health as a carcinogen, mutagen, reproductive toxicant, or other class of toxicant; how the chemical is used in commerce; and what conditions are necessary for the chemical to be used safely. If the quantity of a chemical produced or imported into the EU exceeds 10 tons/year, the business must submit a chemical safety report, which is roughly equivalent to a risk assessment (Figure 9.1), in addition to a technical dossier. The chemical safety report provides in-depth documentation of a chemical's toxicity, if any, to human health and the environment. Criteria for toxicity to human health are incorporated into REACH from earlier EU legislation, and registrants are also required to consider newly recognized forms of toxicity such as the effects of endocrine disruptors. Standardized criteria for environmental toxicity are laid down.

Similar to criteria established by the Canadian Environmental Protection Act, environmental toxicity under REACH includes the characteristics of persistence and bioaccumulation as well as inherent toxicity to ecological receptors (Table 10.1). If the chemical is dangerous to human health or if it is very persistent and very bioaccumulative (see Table 10.1), the chemical safety report must include a detailed exposure assessment and risk characterization that show how the producer or importer is controlling, or intends to control, the risks to human health and the environment at every stage of the chemical's life cycle. While government agencies are primarily responsible for performing risk assessments of industrial chemicals in the United States and Canada, businesses have acquired a much larger role in the European Union under REACH, effectively becoming partners with government in assessing risks associated with the chemicals in their products.

By now the reader will have gathered that the process of registering chemicals under REACH is potentially arduous and burdensome for industry. However, industry was intimately involved in crafting the law, and REACH contains a number of features designed to ease the burden of registration and keep European businesses competitive. One such feature is preregistration of chemicals that are already on the EU market. To preregister a chemical, a producer or importer need submit only basic

TABLE 10.1
Criteria for the Identification of Persistent, Bioaccumulative, and Toxic Substances (PBT substances) and Very Persistent and Very Bioaccumulative Substances (vPvB substances)

Criteria for PBT-Substances	Criteria for vPvB-Substances
Persistent: • Half-life in marine water > 60 days, or • Half-life in fresh- or estuarine water > 40 days, or • Half-life in marine sediment > 180 days, or • Half-life in fresh- or estuarine water sediment > 120 days, or • Half-life in soil > 120 days	**Very Persistent:** • Half-life in marine, fresh-, or estuarine water > 60 days, or • Half-life in marine, fresh-, or estuarine water sediment > 180 days, or • Half-life in soil > 180 days
Bioaccumulative: Bioconcentration factor (BCF) > 2,000	**Very Bioaccumulative** Bioconcentration factor > 5,000
Toxic: • Long-term no-observed-effect concentration (NOEC) for marine or freshwater organisms is less than 0.01 mg/L, or • Substance is classified as carcinogenic (category 1 or 2), mutagenic (category 1 or 2), or toxic for reproduction (category 1, 2, or 3), or • There is other evidence of chronic toxicity, as identified by the classifications: T, R48, or Xn, R48 according to Directive 67/548/EEC	

Source: Summarized from Annex XIII of "Corrigendum to Regulation (EC) 1907/2006 of the European Parliament and of the Council of 18 December 2006 concerning the Registration, Evaluation, Authorization and Restriction of Chemicals (REACH)," *Official Journal of the European Union*, L136, Vol. 50, 29 May 2007. http://eur-lex.europa.eu/JOHtml.do?uri=OJ:L:2007:136:SOM:EN:HTML

information: the name and contact information of the registrant, the name of the chemical, and the approximate tonnage produced or imported. The preregistration process under REACH is roughly equivalent to the TSCA Inventory in the United States, except that tonnage thresholds apply to each business as a whole rather than separately to each manufacturing site within a business, and thresholds are also about five times lower under REACH than under TSCA. Chemicals that were preregistered between June 1 and December 1, 2008 can continue to be manufactured or imported into the EU while registration is phased in over the next several years. Deadlines for final registration of preregistered chemicals are based on tonnage and toxicity: Any chemical that is manufactured or imported in quantities greater than 1,000 tons/year must be registered by December 1, 2010. The same deadline applies to any chemical that is classified on the basis of existing law as a carcinogen, mutagen, or reproductive toxicant and is manufactured or imported in quantities of 1 ton/year or more, or as dangerous to the aquatic environment and manufactured or imported in quantities greater than 100 tons/year. If a chemical is not classified as toxic to human health or aquatic life, it does not have to be registered until June 1, 2013, provided it is manufactured or imported in quantities between 100 and 1,000 tons/year; the registration deadline is June 1, 2018, if the quantities are 1 to 100 tons/year. Preregistration has provided more time for businesses to develop the detailed technical and safety information required for registration of chemicals under REACH. According to the Web site of the European Chemical Agency (ECHA 2009), approximately 143,000 chemicals were preregistered by 65,000 companies by the December 1, 2008, deadline.

In addition to providing more time to comply with registration rules, preregistration has given companies an opportunity to share data. Since each and every company that produces or imports a chemical above a specific annual tonnage is subject to REACH, there is a tremendous potential for companies registering the same chemical to duplicate each other's risk assessments. The cost of such duplication to industry is an issue, but an even greater concern is that duplication could involve unnecessary toxicity testing on vertebrate animals. REACH seeks to reduce animal toxicity testing to the lowest level consistent with good risk-assessment practice, and any company that possesses animal toxicity data is obligated to make its data available to other companies seeking to register the same chemical, in return for fair compensation. Companies are encouraged to share other toxicity data, as well, such as results of in vitro tests of mutagenicity and cellular toxicity or studies of structure-activity relationships (SARs) (Chapter 5). ECHA actively facilitates data sharing by coordinating voluntary networks of businesses called Substance Information Exchange Forums.

European industry is responsible under REACH for assessing the risks associated with the chemicals they use and for safeguarding workers, consumers, and the environment at every stage of a chemical's life cycle, whether the chemical is used by itself or is incorporated into a product. In effect, the premise of REACH is that all chemicals are suspect, and that the burden is on those who profit from chemicals to show how they can be used safely. Industry has significant input into how toxic chemical risk is assessed and managed, but it does not have the final word. That is reserved to government. ECHA, European member states, and the European Commission review registrations and are ultimately responsible, individually and collectively, for identifying, characterizing, and prioritizing substances of very high concern (SVHC). SVHC chemicals are (a) carcinogenic, mutagenic, or toxic to reproduction; (b) persistent, bioaccumulative, and toxic (PBT) or very persistent or very bioaccumulative (vPvB) in the environment (Table 10.1); or are implicated by scientific evidence in other impacts of concern, for example, endocrine disruptors. SVHC chemicals are roughly analogous to chemicals on the Toxic Substances List (Schedule 1) in Canada.

SVHC chemicals are candidates for authorization. A chemical that is subject to authorization is banned from the European Union in any amount unless a specific use is authorized by the European Commission. Authorization is somewhat akin to the "virtual elimination" of a chemical in Canada. The goal under both the European and the Canadian policies is to reduce the quantities of chemicals judged to be extremely harmful to levels consistent with reasonable precaution while preserving enough flexibility to respond to overriding economic concerns, if necessary. REACH turns traditional regulatory priorities upside down. Whereas TSCA requires the U.S. government to prove a chemical is harmful before regulating its use by industry, REACH gives government the right to classify a chemical as harmful and requires industry to show why its use should not be restricted or banned.

Authorization is a four-step process:

1. Substances are identified as SVHC by member countries or by EU agencies such as ECHA, with input from industry and other interested parties.
2. SVHC substances are prioritized by ECHA with input from other government agencies, and decisions are made as to (a) whether a substance will be subject to authorization, (b) which uses do not require authorization (e.g., if they are regulated by other laws), and (c) the "sunset date" after which the chemical may not be used unless it is authorized.
3. Businesses may apply for authorization to use the chemical. An application for authorization must include an analysis of alternatives, and it must offer a substitution plan if an alternative is identified that is feasible and reduces risks.
4. A decision whether or not to grant authorization to use an SVHC chemical is made by the European Commission. Authorization may be granted if a socioeconomic analysis determines that a chemical's overall benefit to society outweighs its risks, or if its use is restricted to a product or process for which there is no viable substitute.

The authorization process epitomizes the new risk-management balance that REACH seeks to strike between human health and the environment, on the one hand, and economic interests, on the other. At the time of this writing, REACH is still a regulatory work in progress, and its final outcome is not yet known. The preregistration phase is complete; producers and importers are proceeding with registration; substance information exchange forums (SIEFs) are forming to facilitate data-sharing on chemical risk; and ECHA has recommended the first seven chemicals and chemical classes for authorization. REACH, like CEPA 1999, is a hopeful new effort to wield the double-edged sword of industrial-scale chemistry in ways that reduce risks to human beings and other species.

STUDY QUESTIONS

1. What is the principal strategy for managing toxic chemical risk? Explain.
2. Describe life-cycle assessment. How is it used to manage risks from toxic chemicals?
3. What is pollution prevention?
4. True or false: Risk management is based solely on scientific risk assessment. Explain.

5. Canada and the European Union have legislated criteria for toxicity based on a chemical's persistence in the environment and its tendency to bioaccumulate. Do you agree that environmentally persistent and bioaccumulative chemicals should be defined as toxic, even when, as is generally the case, they do not produce measurable toxic effects in humans or wildlife? Explain your position.
6. The Toxic Substances Control Act of 1976 (TSCA) was among the first laws to attempt to define and regulate toxic chemicals. The Canadian Environmental Protection Act (CEPA 1999) and the European Union's directive on the Registration, Evaluation, Authorization and Restriction of Chemicals in 2007 (REACH) represent more recent efforts to manage chemical risk.
   a. Name one feature that TSCA, CEPA 1999, and REACH have in common.
   b. Name two provisions in CEPA 1999 and REACH that are missing from TSCA.
   c. Describe one risk-management strategy that is unique to REACH.
7. Describe the registration process for chemicals under REACH.
8. What is meant by "authorization" under REACH?
9. Define the precautionary principle. Discuss the advantages and disadvantages of applying this principle to the management of toxic chemical risk. Give three examples of national governments or international bodies applying the precautionary principle to the management of risk from chemicals.
10. Assume that the U.S. Congress has decided to consider updating TSCA, and you are invited to give expert testimony before the subcommittee responsible for drafting a new law to regulate toxic chemicals in the United States. What would you say to the committee?

ANSWERS TO STUDY QUESTIONS

1. Controlling exposure is the principal strategy for managing toxic chemical risk. Risk consists of two components: the toxicity of a chemical and the conditions of its use. Toxicity is an inherent property and cannot be controlled. Exposure can be reduced by managing the conditions under which a chemical is used.
2. A life-cycle assessment (LCA) examines how a chemical is used from the beginning to the end of an industrial process, from "cradle to grave." An LCA may be performed on a chemical alone or on a chemical as part of a product. The conditions of use are different at different stages of a chemical's life cycle. An LCA opens the door to identifying life-cycle stages where exposure can be reduced. It also facilitates the identification of substitute chemicals that are less toxic.
3. Pollution prevention is a strategy for reducing exposure that can be applied to a life-cycle assessment. Pollution can rarely be prevented altogether, but it can almost always be reduced. For example, pollution might be reduced by using less chemical in the manufacture of a product, by altering the

manufacturing process to reduce the generation of toxic chemical by-products, by recycling the chemical into new products, or by creating a less polluting disposal option. Pollution might also be reduced by substituting a different chemical in the manufacturing process.
4. False. Risk management is a political process in which science has a preferred but by no means exclusive role. Economic considerations, world view, and the psychology of risk perception help shape the laws that govern how a society manages toxic chemical risk.
5. Recent laws in Canada (CEPA 1999) and Europe (REACH) add environmental fate and transport criteria to the traditional definitions of chemical toxicity based on health effects, e.g., carcinogenicity and reproductive toxicity. The new laws reflect growing concern that environmental persistence and bioaccumulation are setting the stage for long-term exposures that available evidence suggests can prove harmful eventually, even if exposures are as yet too low to cause statistically significant toxic effects in many cases. Classifying highly persistent and bioaccumulative chemicals as toxic is consistent with the precautionary principle.
6a. All three laws mandate the compilation of a list of all chemicals in commerce: the TSCA Inventory, the Domestic Substances List (CEPA), and the EU's registration directive (REACH).
  b. Unlike TSCA, the Canadian and European laws provide for sorting chemicals in commerce into toxic and nontoxic categories, followed by the prioritization of toxic chemicals and restrictions on their use. Further, they provide explicit statutory criteria for toxicity. A third difference is that the Canadian and European laws enlarge the definition of toxicity to include environmental persistence and bioaccumulation.
  c. REACH requires manufacturers and importers to take responsibility for submitting a detailed chemical profile ("technical dossier") and risk assessment ("chemical safety report") when they register a chemical with the European Chemical Agency (ECHA). TSCA and CEPA place a greater share of the responsibility on government agencies to characterize risk. Government bears the ultimate responsibility for characterizing and prioritizing risk under REACH, as it does under TSCA and CEPA.
7. Manufacturers and importers are required to register with the European Chemical Agency (ECHA) each and every chemical that they use in quantities of 1 ton/year or more, either alone or in a product. They are required to submit a technical dossier describing each chemical's properties and hazards. If use of the chemical exceeds 10 tons/year, they are required to submit a chemical safety report with detailed information on chemical risk and how exposure is controlled. To ease the burden of compliance, companies were given an opportunity between June 1 and December 1, 2008, to pre-register chemicals that were already in commerce in the European Union by reporting the name of the chemical, the company, and the tonnage to ECHA. Deadlines for registration extend from 2010 to 2018, depending on the preregistered chemical's toxicity and tonnage. In addition to extended deadlines, REACH facilitates the registration process and also reduces the

sacrifice of vertebrate animals by requiring companies to share the results of animal toxicity tests and encouraging them to share other types of data, as well. ECHA coordinates voluntary substance information exchange forums (SIEFs) that companies can join to share data required to register chemicals.
8. The use of chemicals that are determined by ECHA to be substances of very high concern, also referred to as SVHC chemicals, may be restricted or banned. In the language of REACH, an SVHC chemical may become subject to authorization. Once a chemical is subject to authorization, its use in any industrial process must be specifically authorized, or approved, by the European Commission.
9. The precautionary principle is sometimes compared to the Hippocratic oath, "First, do no harm." It recognizes both the limits of human knowledge as well as the human capacity, individually and collectively, to do harm, even when acting with good intentions. Applied to the management of toxic chemical risk, the precautionary principle advises that it may be prudent to seek to avoid harm by reducing exposure, even though the evidence is inconclusive and the scientific uncertainty of the risk assessment is relatively large. Examples of the application of the precautionary principle include: the recent Canadian ban on the importation of baby bottles containing bisphenol A; the inclusion of environmental persistence and bioaccumulation as criteria for toxicity in new environmental laws in Canada and the European Union; and the 1997 Kyoto Protocol and other efforts to reduce greenhouse gas emissions and reduce the impacts of global warming. The 1987 Montreal Protocol phasing out the use of chlorofluorocarbons to protect stratospheric ozone was based on extensive scientific evidence and, therefore, although a landmark achievement, does not qualify as an application of the precautionary principle.
10. Reader's choice.

REFERENCES

Cornell University Law School, Legal Information Institute. U.S. Code collection, Title 15, Chapter 53: Toxic Substances Control. http://www4.law.cornell.edu/uscode/15/ch53.html (accessed July 19, 2009).

ECHA. REACH guidance. European Chemical Agency. http://guidance.echa.europa.eu/index_en.htm (accessed July 19, 2009).

Government of Canada. Chemical substances portal. http://www.chemicalsubstanceschimiques.gc.ca/en/index.html

Houghton, J. 2004. *Global warming: The complete briefing*. Cambridge: Cambridge University Press.

Kolbert, E. 2009. The sixth extinction? The earth's species in peril. *The New Yorker*, May 25, 52–63.

Schapiro, M. 2007. *Exposed: The toxic chemistry of everyday products and what's at stake for American power*. White River Junction, VT: Chelsea Green Publishing.

U.S. EPA. TSCA statutes, regulations and enforcement. Environmental Protection Agency. http://www.epa.gov/compliance/civil/tsca/tscaenfstatreq.html (accessed July 19, 2009).

USGS, Sound Waves Monthly Newsletter. Coral-reef builders vulnerable to ocean acidification. http://soundwaves.usgs.gov/2008/03research.html
Vaccari, D. A. 2009. Phosphorus: A looming crisis. *Scientific American* 300 (6), 54–59.

SUGGESTED READING

Baker, N. 2008. *The body toxic*. New York: North Point Press.

Index

A

Acetylaminofluorene, 102
Acute lethality, 71–73
Adenosine triphosphate, 123
Advective transport, 13, 19–21
Afferent arteriole, kidney, 99
Air quality index, 144–145
Alcohol
    acute oral lethal dose, 72
    cancer deaths from, 128
    lay *vs.* expert risk perceptions, 136
Aldrin, 102
Ames test, 87–88, 90–91, 107
Aminopyrine, 102
Ammonia, acute oral lethal dose, 72
Amobarbital, 102
Amphetamine, 102
Aniline, 102
Antarctic food web, 26
AQI. *See* Air quality index
Aquifers, 28–29
Arfarin, 102
Arsenic, 19, 59, 63, 71–72, 137–140, 147–148
    acute oral lethal dose, 72
Asbestos, cancer from, 126
Aspirin, acute oral lethal dose, 72
Atomic bomb attack on Hiroshima, Japan, 60
ATP. *See* Adenosine triphosphate

B

Bacteria for screening chemicals, 87–88
Basal body, 96
Benzene
    aqueous solubility, 16
    bioconcentration factors, 25
    vapor pressure, Henry's law constant, 15
Benzene hexachloride delta isomer, acute oral lethal dose, 72
Benzphetamine, 102
Benzpyrene, 102
Bioaccumulation, 24–27
Bioavailability, 94–95
Bioconcentration, 24–27
Biogeochemical cycles, 27
Biomagnification, 24–27
Biomonitoring, 141–142
Biotransformation, 101–107

Birds, decreased reproduction, dichlorodiphenyltrichloroethane, 3
Birth defects, thalidomide, 3–4
Blue-green algae, toxicity testing, 68
Bobwhite, toxicity testing, 68
Bodily defenses, 93–118
    basal body, 96
    bioavailability, 94–95
    biotransformation, 101–107
    carbohydrate, 98
    cell membrane, 95–98
    centrioles, 96
    cholesterol molecules, 98
    chromatin, 96
    cilia, 96
    elimination, 99–101
    eukaryotic cell with organelles, 96
    excretion, 99–101
    exposure, 94–95
    fibrous proteins, 98
    flagellum, 96
    fluidity of cell membrane, 98
    globular protein, 98
    glucuronidation, 104
    glycolipid, 98
    Golgi apparatus, 96
    kidneys, elimination by, 98–99
    liver structure, 105
    lysosome, 96
    microtubule, 96
    microvilli, 96
    mitochondrion, 96
    nuclear envelope, 96
    nucleolus, 96
    nucleus, 96
    phospholipid head, 98
    phospholipid molecules, double layer, 98
    phospholipid tail, 98
    repeated-dose exposure, kinetics of, 112–114
    ribosomes, 96
    rough endoplasmic reticulum, 96
    secretory vesicle, 96
    single-dose exposure, kinetics of, 107–112
    smooth endoplasmic reticulum, 96
    Tylenol overdose, 106
    weak acids, bases, 99–101
Botulinus toxin, acute oral lethal dose, 72
Brachen fern, 127
Brockovich, Erin, 61–62

Butylated hydroxytoluene, chemical structure, 130

C

Cadmium
  bioconcentration factors, 25
  cancer from, 126
Caffeine, 102
CAFO. *See* Concentrated animal feedlot operation
Canadian Environmental Protection Act, 175–177
Cancer, 6, 59–60, 62–63, 68–70, 124–129, 145–148, 151–154
  chronic daily intake, 147
  deaths from environmental factors, 128
  hazard index for noncancer health effects, 147
  industrial chemicals causing, 126
  risk calculations, 147–148
Carbohydrate, 98
Carbon monoxide, 59, 119, 121–122, 124, 133, 144
Carbon tetrachloride, 103
Carcinogenicity testing, 73–74
Case-control studies, 62–63
Catfish, toxicity testing, 68
Causation proof, 58–60
Cell membrane, 95–98
  fluidity, 98
Centrioles, 96
CEPA. *See* Canadian Environmental Protection Act
CERCLA. *See* Comprehensive Environmental Response, Compensation and Liability Act
Chemical classes, classification based on, 2
Chemical transformation, 21–24
Chlorphentermine, 102
Chlorpromazine, 102
Chlorpropamide, 102
Cholesterol molecules, 98
Chromatin, 96
Chronic daily intakes, 138–141, 147
Chronic toxicity, 73–74
Cigarette smoke, 59–60, 126
Cigarette smoking, cancer from, 126
Cilia, 96
Cimetidine, 102
Classification schemes, 2
  based on chemical classes, 2
  based on exposure pathways, 2
  based on mechanisms of toxicity, 2
  based on sources of toxic chemicals, 2
  based on usage of chemicals, 2
Clean Air Act, 22, 54, 137, 144, 153, 171
Coal tar, cancer from, 126
Codeine, 102

Cohort studies, 62–63
Comprehensive Environmental Response, Compensation and Liability Act, 54, 61, 136–137
Concentrated animal feedlot operation, 13
Consumer Product Safety Act, 54
Costs of toxic chemicals, 169–171
Costs of toxicity testing, 70–71
Crayfish, toxicity testing, 68
Crotolaria, 127
Cultured mammalian cells, genetic toxicity, 88–89
Cumulative dose–effect curve, features of, 43
Cumulative toxicity, 42–43
Cyanide, 59, 123, 133

D

DDE. *See* Dichlorodiphenyldichloroethylene
DDT. *See* Dichlorodiphenyltrichloroethane
Defenses, bodily, 93–118. *See also* Bodily defenses
Dermal route of exposure, 69
Design of toxicity test, 68–71
Diazepam, 102
Dichlorodiphenyldichloroethylene, 3
Dichlorodiphenyltrichloroethane, 3
  acute oral lethal dose, 72
  decreased reproduction in birds, 3
Dieldrin, acute oral lethal dose, 72
Diethylnitrosamine, structure, 126
Diffusive transport, 13
Digitoxin, 102
Dimethylamino-azobenzene, structure, 126
Dimethylbenz(a)-anthracene, structure, 126
Dioxin
  acute oral lethal dose, 72
  aqueous solubility, 16
  bioconcentration factors, 25
  vapor pressure, Henry's law constant, 15
Disease mechanisms, 119–134
  adenosine triphosphate, 123
  alcohol, cancer deaths from, 128
  butylated hydroxytoluene, chemical structure, 130
  cancer, 124–132
  cancer deaths from environmental factors, 128
  developmental toxicity, 124
  diethylnitrosamine, structure, 126
  dimethylamino-azobenzene, structure, 126
  dimethylbenz(a)-anthracene, structure, 126
  estradiol benzoate, chemical structure, 130
  ethionine, structure, 126
  griseofulvin, structure, 127
  hydroxyxanthine, structure, 126
  methylcholanthrene, structure, 126

Index

mitomycin C, structure, 127
N-methy-N-formylhydrazine, structure, 127
naphthylamine, structure, 126
naturally occurring chemicals causing cancer, 124
nitroquinoline-1-oxide, structure, 126
noncancer health effects, 119–124
organ toxicity, 120–124
phenobarbital, chemical structure, 130
promoting agents chemical structures, 130
pyrrolizidine alkaloids, structure, 127
quercetin, structure, 127
saccharin, chemical structure, 130
safrole, structure, 127
stages in internal development of disease, 121
target molecules, toxicant molecules, interactions of, 125
tetradecanoyl phorbol acetate, chemical structure, 130
thiourea, structure, 126
tobacco, cancer deaths from, 128
urethan, structure, 126
Dissolution, 16–17
Domestic Substances List, 175
Dose, in toxicity tests, 69–70, 77
Dose–effect, 37–52
   ethical dilemmas, 37–38
   graded dose–effect relationship, 43–48
   protection of public health, 37–38
   quantal dose–effect relationship, 38–43
   toxicity, preliminary investigations, 38
DSL. *See* Domestic Substances List
Dye intermediates, cancer in bladder, 126

E

East Woburn, Massachusetts, leukemia incidence, 61
Ecological risk assessment, 157–168
   environmental impact statement, 166
   Environmental Protection Agency process, 158–166
   framework, 157–158
Ecosystems, 27
ECVAM. *See* European Centre for Validation of Alternative Methods
Edible false morel mushroom, N-methy-N-formylhydrazine, 127
Effects analysis, 142–148
   air quality index, 144–145
   Clean Air Act, 144
   margin of exposure, 145
   margin of safety, 145
   National Oceanic and Atmospheric Administration, 144
   no observed adverse effects level, 142–143
   noncancer health effects, 142–145
   therapeutic index, to evaluate safety of drugs, 145
Efficacy of toxicity testing, 82–84
Eggshell thinning, dichlorodiphenyltrichloroethane, 3
Electric power lay *vs.* expert risk perception, 136
Elimination, 99–101
Endpoint, in toxicity tests, 70
Environmental impact statement, 166
Environmental pathways, 13–35
   advective transport, 13, 19–21
   aquifers, 28–29
   bioaccumulation, 24–27
   bioconcentration, 24–27
   bioconcentration factors, 25
   biogeochemical cycles, 27
   biomagnification, 24–27
   chemical transformation, 21–24
   concentrated animal feedlot operation, 13
   diffusive transport, 13
   ecosystems, 27
   exposure assessment, 29–31
   food web, Antarctic, 26
   Henry's law constants, 15
   hydrologic cycle, 27–29
   management of exposure, 29–31
   partitioning, 14–19
   partitioning adsorption, 18–19
   persistent organic pollutants, 14
   physical states of water, 29
   processes governing movement of chemicals, 13
   vapor pressures, 15
Environmental Protection Agency, 54, 119, 136–142, 144–150, 157–159, 161–163, 174–175, 184
   adaptive management, 165–166
   analysis phase, 162–165
   ecosystem, receptor characteristics, measures of, 163
   effects, measures of, 163
   exposure, measures of, 163
   management goals, 159–160, 163
   planning, 159–160
   problem formulation, 160–162
   process, 158–166
   risk characterization, 165
   risk communication, 165–166
   standard default exposure factors, 139
EPA. *See* Environmental Protection Agency
Epidemiological study design, 60–63
Estradiol benzoate, chemical structure, 130
Ethical dilemmas, 37–38
Ethinyl estradiol, 102
Ethionine, structure, 126
Ethylmorphine, 102
Eukaryotic cell with organelles, 96

European Centre for Validation of Alternative Methods, 86
Evaporation, 14–16
Excretion, 99–101
Exposure, 94–95
Exposure analysis, 138–142
    biomonitoring, 141–142
    chronic daily intake, 138–141
    standard default exposure factors, Environmental Protection Agency, 139
Exposure assessment, 29–31
Exposure levels, toxic effect frequencies, 78–79
Exposure pathways, classification based on, 2
Eye irritation test, in personal-care product screening, 85

F

Fathead minnow, toxicity testing, 68
FDA. *See* Food and Drug Administration
Federal Water Pollution Control Act, 54
Fibrous proteins, 98
FIFRA. *See* Fungicides, Insecticides, and Rodenticides Act
Flagellum, 96
Flavonoid, 127
Fluidity of cell membrane, 98
Food and Drug Administration, 3, 54, 67, 150, 154
Food and Drug Administration Modernization Act, 54
Food Quality Protection Act, 54
Food web, Antarctic, 26
Foods, Drugs, and Cosmetics Acts, 54
Fossil fuel burning, 29, 55, 137, 170, 184
FQPA. *See* Food Quality Protection Act
Fungicides, Insecticides, and Rodenticides Act, 54

G

Global warming, 29, 55, 137, 170, 184
Globular protein, 98
Glucuronidation, 104
Glutethimide, 102
Glycolipid, 98
Goldfish, toxicity testing, 68
Golgi apparatus, 96
Graded dose–effect relationship, 43–48
Green algae, toxicity testing, 68
Greenhouse gases, 29, 55, 137, 167, 170, 184
Griseofulvin, structure, 127

H

Handguns lay *vs.* expert risk perception, 136

Hazard identification, 136–138
Hazard index for noncancer health effects, 147
Health risk assessment, 135–155
    cancer risk, 145–148
    Comprehensive Environmental Response, Compensation and Liability Act, 136–137
    effects analysis, 142–148
    exposure analysis, 138–142
    hazard identification, 136–138
    process of risk assessment, 136
    risk characterization, 148–149
Heart attack risk, rofecoxib, 4
Heliotropium, 127
Henry's law, 15, 171
    constants, 15
Heptachlor, acute oral lethal dose, 72
HI. *See* Hazard index
Hydrologic cycle, 27–29
Hydroxyxanthine, structure, 126

I

Ibuprofen, 102
ICCVAM. *See* Interagency Coordinating Committee for Validation of Alternative Methods
Impact statement, 166
Incremental toxicity, 41–42
Individual organism *vs.* population investigations, 84–85
Information derived from toxicity tests, 78–84
Inhalation route of exposure, 69
Interagency Coordinating Committee for Validation of Alternative Methods, 86

K

Kidneys, elimination by, 98–99

L

LCA. *See* Life-cycle assessment
LD. *See* Median lethal dose
Lead, 2–3, 7–8, 24, 131–132, 160–162, 165, 173–174
Life-cycle assessment, 174
Limitation of use of toxic chemicals, 173–174
Lindane
    acute oral lethal dose, 72
    aqueous solubility, 16
    bioconcentration factors, 25
    vapor pressure, Henry's law constant, 15
Liver structure, 105
Log-dose effect curve, 82
Lung cancer, 59–60
Lung disease, 59–60, 126

Index

Lysosome, 96

M

Malathion
  acute oral lethal dose, 72
  aqueous solubility, 16
  bioconcentration factors, 25
  vapor pressure, Henry's law constant, 15
Management of exposure, 29–31
Margin of exposure, 145
Margin of safety, 145
Marine Protection Research, Sanctuaries Act, 54
MATC. See Maximum acceptable toxicant concentration
Maximum acceptable toxicant concentration, 75
Maximum tolerated dose, 74
Mayflies, toxicity testing, 68
Mechanisms of toxicity, classification based on, 2
Median lethal dose, 71
Meprobamate, 102
Mercury, 13–15, 19, 63, 68, 137, 150, 154, 173
Methaqualone, 102
Methitural, 102
Methoxychlor, acute oral lethal dose, 72
Methylcholanthrene, structure, 126
Methylthiopurine, 102
Mice, toxicity testing, 68
Microvilli, 96
Midges, toxicity testing, 68
Mitochondrion, 96
Mitomycin C, structure, 127
MOE. See Margin of exposure
Morphine, 102
MOS. See Margin of safety
Motor vehicles lay vs. expert risk perception, 136
Motorcycles lay vs. expert risk perception, 136
Movement of chemicals, processes governing, 13
Mphetamine, 102
MTD. See Maximum tolerated dose

N

N-methy-N-formylhydrazine, structure, 127
Naphthalene, 102
Naphthylamine, structure, 126
National Environmental Policy Act, 54
National Oceanic and Atmospheric Administration, 144
Naturally occurring chemicals causing cancer, 124
NEPA. See National Environmental Policy Act
Nicotine, 102
  acute oral lethal dose, 72
Nitroanisole, 102
Nitrophenol
  aqueous solubility, 16
  bioconcentration factors, 25
  vapor pressure, Henry's law constant, 15
Nitroquinoline-1-oxide, structure, 126
No observed adverse effect level, 79–80, 142–143
NOAA. See National Oceanic and Atmospheric Administration
NOAEL. See No observed adverse effect level
Noncancer health effects, 119–124
Nonmammalian species, toxicity test design in, 75–76
Nuclear envelope, 96
Nucleolus, 96
Null hypothesis, 57–58

O

Occupational Safety and Health Administration, 54, 67
Ocean Radioactive Dumping Ban Act, 54
Oral route of exposure, 69
Organisms used in toxicity tests, 68
OSHA. See Occupational Safety and Health Administration
Overview of science of toxicology, 1–4

P

Parathion, 103
Partitioning, 14–19. See also Diffusive transport
  adsorption, 18–19
  dissolution, 16–17
  evaporation, 14–16
  volatilization, 17–18
PBBs. See Polybrominated biphenyls
PCBs. See Polychlorinated biphenyls
PCDFs. See Polychlorinated dibenzofurans
Pentachlorophenol
  aqueous solubility, 16
  bioconcentration factors, 25
  vapor pressure, Henry's law constant, 15
Pentobarbital, 102
Peritubular capillary, kidney, 99
Persistent organic pollutants, 14
Personal-care product screening, 85–86
  eye irritation test, 85
  skin irritation test, 85
  skin sensitization test, 85–86
Pesticides lay vs. expert risk perception, 136
Phenobarbital, 102
  chemical structure, 130
Phenol
  aqueous solubility, 16
  bioconcentration factors, 25
  vapor pressure, Henry's law constant, 15
Phenylbutazone, 102
Phenytoin, 102
Phospholipid head, 98

Phospholipid molecules, double layer, 98
Phospholipid tail, 98
Physical states of water, 29
Pipe smoking, cancer risk, 126
Planaria, toxicity testing, 68
Poison, usage of term, 2
Political process, navigation of, 172
Polybrominated biphenyls, acute oral lethal dose, 72
Polychlorinated biphenyls, 59, 173
   toxicity, 60
Polychlorinated dibenzofurans, 59
POPs. *See* Persistent organic pollutants
Populations at risk, 53–66
   atomic bomb attack on Hiroshima, Japan, 60
   Brockovich, Erin, 61–62
   cancer, 59–60
   case-control studies, 62–63
   causation proof, 58–60
   Clean Air Act 1970, 22, 54, 137, 144, 153, 171
   cohort studies, 62–63
   Comprehensive Environmental Response, Compensation and Liability Act, 61
   Consumer Product Safety Act, 54
   East Woburn, Massachusetts, leukemia incidence, 61
   Environmental Protection Agency, 54, 136–142, 144–150, 157–159, 161–163, 165, 174–175, 184
   epidemiological study design, 60–63
   Federal Water Pollution Control Act 1972, 54
   Food and Drug Administration Modernization Act, 54
   Food Quality Protection Act, 54
   Foods, Drugs, and Cosmetics Acts, 54
   Fungicides, Insecticides, and Rodenticides Act, 54
   legal issues, 53–55
   level I epidemiological study, 63
   level II epidemiological study, 63
   level III epidemiological study, 63
   lung cancer, 59–60
   Marine Protection Research, Sanctuaries Act, 54
   National Environmental Policy Act, 54
   null hypothesis, 57–58
   Occupational Safety and Health Administration, 54
   Ocean Radioactive Dumping Ban Act, 54
   polychlorinated biphenyls, 59
   polychlorinated dibenzofurans, 59
   Resource Conservation and Recovery Act, 54
   Safe Drinking Water Act 1974, 54
   statistical power, 57–58
   tobacco cigarette smoke, lung disease from, 59–60
   Toxic Substances Control Act, 53–55, 149, 173–175, 177–179, 181–184
   toxic tort lawsuits, 61–62
   transport of hazardous materials, 54
Potency, in toxicity tests, 82–84
Probit plot, 76–78
Propranolol, 102
Protection of public health, 37–38
Pyrrolizidine alkaloids, 127
   structure, 127

Q

Quantal dose–effect relationship, 38–43
   cumulative dose–effect curve, features of, 43
   cumulative toxicity, 42–43
   Gaussian, distributions, 41
   incremental toxicity, 41–42
Quercetin, structure, 127

R

Radioactive watch dial dyes, cancer in bone, 126
Rainbow trout, toxicity testing, 68
Rate of increase in chemical disease frequency, dose and, 81–82
Rats, toxicity testing, 68
RCRA. *See* Resource Conservation and Recovery Act
REACH. *See* Registration, Evaluation, Authorization and Restriction of Chemicals
Reducing use of animals in toxicity tests, 86–89
Registration, Evaluation, Authorization and Restriction of Chemicals, 177–181
Renal tubule, kidney, 99
Repeated-dose exposure, kinetics of, 112–114
Reproductive toxicity testing, 70, 74–75
Resource Conservation and Recovery Act, 54, 174
Ribosomes, 96
Rice oil disease, 59
Ring-necked pheasant, toxicity testing, 68
Risk characterization, 148–149
   uncertainty, 148–149
   weight of evidence, 149
Risk management, 169–185
   Canadian Environmental Protection Act, 175–177
   controlling exposure, 171–172
   costs of toxic chemicals, 169–171
   damage to environment, 170–171
   Domestic Substances List, 175
   environmental laws, 174–181
   fossil fuel burning, 29, 55, 137, 170, 184
   global economy, 29, 55, 137, 170, 184
   global warming, 29, 55, 137, 170, 184

Index

greenhouse gases, 29, 55, 137, 170, 184
Henry's law, 171
life-cycle assessment, 174
limiting use of toxic chemicals, 173–174
navigating political process, 172
polychlorinated biphenyls, 173
Registration, Evaluation, Authorization and Restriction of Chemicals, 177–181
sorting chemicals, 172–173
Toxic Substances Control Act, 173–175
uncertainty, dealing with, 172
Waste in Electrical and Electronic Equipment directive, 174
Rofecoxib, risk of heart attack, stroke, 4
Rough endoplasmic reticulum, 96
Route of exposure, 69

S

Saccharin, chemical structure, 130
Safe Drinking Water Act, 54
Safrole, 127
structure, 127
Salt, acute oral lethal dose, 72
Science, toxic chemical risk as, 1–12
Secobarbital, 102
Secretory vesicle, 96
Senecio, 127
Single cell toxicity testing, 86–87
Single-dose exposure, kinetics of, 107–112
Skin irritation test, in personal-care product screening, 85
Skin sensitization test, in personal-care product screening, 85–86
Smoking, 59–60, 126
of cancer of lung, 126
lay vs. expert risk perception, 136
Smooth endoplasmic reticulum, 96
Social discourse, toxic chemical risk as, 1–12
Soot, cancer risk, 126
Specified toxic effect, in toxicity testing, 70
Stages in internal development of disease, 121
Standard default exposure factors, Environmental Protection Agency, 139
Statistical power, 57–58
in toxicity tests, 70–71
Stroke risk, rofecoxib, 4
Structure, activity, relationship between, 89
Strychnine, acute oral lethal dose, 72
Subchronic toxicity testing, 73
Sugar, acute oral lethal dose, 72
Sunlight, skin cancer, 126
Superfund. See Comprehensive Environmental Response, Compensation and Liability Act
Surgery, lay vs. expert risk perception, 136
Swimming, lay vs. expert risk perception, 136

T

Target molecules, toxicant molecules, interactions of, 125
TCDD. See Dioxin
Tetradecanoyl phorbol acetate, chemical structure, 130
Thalidomide, 3
missing limbs in newborns, 3
teratogenicity, 3–4
Theophylline, 102
Therapeutic index, to evaluate safety of drugs, 145
Thiopental, 103
Thioridazine, 102
Thiourea, structure, 126
Threshold of toxicity, 79–81
Time frame, in toxicity testing, 69–70
Tobacco, 59–60, 126
cancer deaths, 128
cancer of lung, 126
lung disease, 59–60, 126
Tobacco juices, cancer in oral cavity, 126
Toxic Substances Control Act, 53–55, 149, 173–175, 177–179, 181–184
Toxic tort lawsuits, 61–62
Toxicity testing, 67–92
acute lethality, 71–73
Ames test, 87–88, 90–91, 107
bacteria for screening chemicals, 87–88
chronic toxicity, 73–74
cost of, 70–71
cultured mammalian cells, genetic toxicity, 88–89
dermal route of exposure, 69
descriptions, 71–76
design of toxicity test, 68–71
dose, 69–70, 77
efficacy, 82–84
endpoint, 70
European Centre for Validation of Alternative Methods, 86
exposure levels, toxic effect frequencies, 78–79
Food and Drug Administration, 3, 54, 67, 150, 154
individual organism vs. population investigations, 84–85
information derived from, 78–84
inhalation route of exposure, 69
Interagency Coordinating Committee for Validation of Alternative Methods, 86
log-dose effect curve, 82
maximum acceptable toxicant concentration, 75
maximum tolerated dose, 74
median lethal dose, 71

no observed adverse effect level, 79–80
nonmammalian species, toxicity test design
 in, 75–76
Occupational Safety and Health
 Administration, 54, 67
oral route of exposure, 69
organisms used, 68
personal-care product screening, 85–86
potency, 82–84
probit plot, 76–78
rate of increase in chemical disease
 frequency, dose and, 81–82
reducing use of animals in, 86–89
reproductive toxicity, 70
reproductive toxicity testing, 74–75
route of exposure, 69
single cell toxicity testing, 86–87
statistical power, 70–71
structure, activity, relationship between, 89
subchronic toxicity testing, 73
threshold of toxicity, 79–81
time frame, 69–70
U.S. Environmental Protection Agency, 54,
 136–142, 144–150, 157–159, 161–163,
 175, 184
TPA. See Tetradecanoyl phorbol acetate
Transport of hazardous materials, 54
Trichloroethylene
 aqueous solubility, 16
 bioconcentration factors, 25
 vapor pressure, Henry's law constant, 15
TSCA. See Toxic Substances Control Act
Tubular reabsorption, kidney, 99
Tylenol overdose, 106

U

Uncertainty in risk characterization, 148–149
Urethan, structure, 126

V

Vapor pressures, 15
Vinegar, acute oral lethal dose, 72
Vinyl chloride
 aqueous solubility, 16
 bioconcentration factors, 25
 vapor pressure, Henry's law constant, 15
Vioxx. See Rofecoxib
Volatilization, 17–18

W

Warming, global, 29, 55, 137, 170, 184
Waste in Electrical and Electronic Equipment
 directive, 174
Weak acids, bases, 99–101
WEEE directive. See Waste in Electrical and
 Electronic Equipment
Weight of evidence in risk characterization, 149

X

X-rays
 lay vs. expert risk perception, 136
 skin cancer, 126